SpringerBriefs in Molecular Science

History of Chemistry

Series editor
Seth C. Rasmussen, Fargo, USA

For further volumes:
http://www.springer.com/series/10127

Seth C. Rasmussen

The Quest for Aqua Vitae

The History and Chemistry of Alcohol
from Antiquity to the Middle Ages

Seth C. Rasmussen
Department of Chemistry and Biochemistry
North Dakota State University
Fargo, ND
USA

ISSN 2212-991X
ISBN 978-3-319-06301-0   ISBN 978-3-319-06302-7 (eBook)
DOI 10.1007/978-3-319-06302-7
Springer Cham Heidelberg New York Dordrecht London

Library of Congress Control Number: 2014936430

© The Author(s) 2014
This work is subject to copyright. All rights are reserved by the Publisher, whether the whole or part of the material is concerned, specifically the rights of translation, reprinting, reuse of illustrations, recitation, broadcasting, reproduction on microfilms or in any other physical way, and transmission or information storage and retrieval, electronic adaptation, computer software, or by similar or dissimilar methodology now known or hereafter developed. Exempted from this legal reservation are brief excerpts in connection with reviews or scholarly analysis or material supplied specifically for the purpose of being entered and executed on a computer system, for exclusive use by the purchaser of the work. Duplication of this publication or parts thereof is permitted only under the provisions of the Copyright Law of the Publisher's location, in its current version, and permission for use must always be obtained from Springer. Permissions for use may be obtained through RightsLink at the Copyright Clearance Center. Violations are liable to prosecution under the respective Copyright Law.
The use of general descriptive names, registered names, trademarks, service marks, etc. in this publication does not imply, even in the absence of a specific statement, that such names are exempt from the relevant protective laws and regulations and therefore free for general use.
While the advice and information in this book are believed to be true and accurate at the date of publication, neither the authors nor the editors nor the publisher can accept any legal responsibility for any errors or omissions that may be made. The publisher makes no warranty, express or implied, with respect to the material contained herein.

Printed on acid-free paper

Springer is part of Springer Science+Business Media (www.springer.com)

Acknowledgments

I would first and foremost like to thank the National Science Foundation (CHE-0132886) for initial support of this research and the Department of Chemistry and Biochemistry of North Dakota State University for supporting my continuing efforts in the history of chemistry.

The historical work included in the current volume began as an offshoot of my interest in the early introduction and development of chemical glassware in the thirteenth century, including its application to new forms of distillation apparatus. As I had previously undertaken research on the isolation of alcohol via the distillation of wine, I agreed to present an overview of the early history of alcohol as a HIST Tutorial at the 244th National Meeting of the American Chemical Society (ACS) in 2013 as part of programming for the Division of the History of Chemistry (HIST). In the preparation for that presentation, I quickly came to realize that this was historically a pretty murky topic with a number of different viewpoints and biases and few publications that attempted to give a clear view of the big picture of the historical progression over the large timespan involved. As a result, the presentation ultimately given at the 244th ACS Meeting felt like a good start, but a less complete picture of the history involved, which then led to additional research and preparation of the current SpringerBrief volume.

As part of the preparation of this volume, I would like to thank Dr. Valarie Steele of the British Museum for helpful discussions on the chemical analysis of organic residues and biomarkers for the detection of beer and wine, as well as Dr. Stuart Haring of North Dakota State University (NDSU) for discussions on yeast biochemistry and Dr. Erika Offerdahl of NDSU for discussions of general sugar biochemistry and fermentation. In addition, I would like to acknowledge the Interlibrary Loan Department of NDSU, which went out of its way to track down many elusive and somewhat obscure sources. Lastly, I would like to thank the following current and former members of my materials research group at NDSU for reading various drafts of this manuscript and providing critical feedback: Dr. Christopher Heth, Dr. Michael Mulholland, Kristine Konkol, Casey McCausland, Eric Uzelac, Ryan Schwiderski, and Trent Anderson.

Finally, and perhaps most importantly, I must express my continued thanks to Elizabeth Hawkins at Springer, without whom this growing series of historical volumes would not have become a reality.

Contents

1	**Introduction**	1
	1.1 Origin of the Term Alcohol	2
	1.2 Origin of Ethanol Production via Fermentation	7
	1.3 Scope of Current Volume	9
	References	10
2	**Earliest Fermented Beverages**	13
	2.1 Fermentation	13
	2.2 Mead	18
	2.3 Date Wine	22
	2.4 Palm Wine	24
	References	25
3	**Beer**	29
	3.1 Beer versus Bread	30
	3.2 Barley Beer	32
	3.3 Beer versus Wine	34
	3.4 Beer in Mesopotamia	36
	3.5 Beer in Egypt	42
	References	47
4	**Grape Wine**	49
	4.1 Chemical Archaeological Studies	51
	4.2 Viniculture	57
	4.3 Wine Production	62
	References	67
5	**Fermented Milk**	71
	5.1 Kefir	72
	5.2 Kumis	75
	References	76

6	**Distillation and the Isolation of Alcohol**	79
	6.1 A Brief History of Distillation Methods	79
	6.2 Distillation of Wine	88
	References	92
7	**Early Chemical and Medical Applications of Alcohol**	95
	7.1 Chemical and Medical Uses of Fermented Beverages	95
	7.2 Early Chemical Applications of Alcohol	98
	7.3 Early Medical Applications of Alcohol	99
	References	103
About the Author		107
Index		109

Chapter 1
Introduction

Ethyl alcohol, or ethanol, is one of the most ubiquitous chemical compounds in the history of the chemical sciences. Its most common use is as a quite versatile solvent, where it represents one of the very first nonaqueous solvents and most certainly the first polar solvent in this class. Not only is it miscible with both water and wide variety of other organic solvents, but it can solubilize a broad range of analytes. This includes most salts and other water-soluble substances, as well as a great many organic materials not soluble in water, such as fats, resins, and essential oils. As such, it still remains one of the most common chemical media for a wide range of solution-based chemical processes. Beyond its ability to dissolve a large number of chemical species, the antibacterial and antifungal properties of ethanol provide a medium for the preservation of organic matter, as well as an effective disinfectant in medical applications.

Of course, the presence and uses of ethanol go far beyond chemical and medical applications, and as the psychoactive component of fermented beverages it plays a central role in the history of society in general. In fact, ethanol via such fermented beverages is one of the oldest recreational drugs known, and is still the most widely accepted of such drugs in most cultures.

More recently, the flammable nature of ethanol has resulted in its application as a fuel. While such early uses were limited to sources of heating and lighting, ethanol has now become a common fuel or fuel additive for the combustion engine. In this latter application, the ability to produce ethanol from the fermentation of biomass provides the attractive promise of renewable alternatives to petroleum fuels.

Due to its importance and widespread use, the documentation of its history has considerable merit. Not only does this allow us to better understand the sequence of events that led to its initial production via fermentation, but also the confluence of events that ultimately resulted in its successful isolation by distillation. In turn, the history and application of this important chemical species provides critical insight into its overall impact on the progress of chemical practices.

1.1 Origin of the Term Alcohol

Alcohols, along with ethers, may be regarded as derivatives of water, in which one or both of the hydrogen atoms have been replaced by carbon. During the initial classification of organic compounds in the 1800s, these species were classified as belonging to the *'water type'* (Fig. 1.1), as established in 1850 by Alexander Williamson (1824–1904)[1] (Fig. 1.2) [1–4]. As a chemical class, alcohols are abundant in nature and simple alcohols are important chemical species for applications as solvents, fuels, and synthetic intermediates.

Prior to the 1830s, however, the term alcohol referred not to a class of organic compounds, but simply to ethyl alcohol (CH_3CH_2OH). In fact, in 1810, John Dalton gave a symbol representing alcohol as a single compound in his *New System of Chemical Philosophy* (Fig. 1.3) [5]. The use and meaning of the term alcohol then changed in 1834 with the discovery of methyl alcohol (CH_3OH) by Jean Baptiste Dumas (1800–1884)[2] and Eugène Peligot (1811–1890)[3] (Fig. 1.4)

---

[1] Alexander William Williamson (1824–1904) was born to Scottish parents in the London borough of Wandsworth, England [2–4]. He entered the University of Heidelberg in 1841 to fulfill his father's wish that he study medicine [3], but soon gave up medical studies to pursue chemistry under Leopold Gmelin (1788–1853) [2–4]. He moved to Giessen in 1844 to complete his chemical education under Justus Liebig (1803–1873) [2–4] before finally moving to Paris in 1846 where he studied mathematics under Augusta Comte (1798–1857) at the École Polytechnique [3, 4]. On the basis of important research on hypochlorous acid and on Prussian blue, he then succeeded George Fownes (1815–1849) as the professor of analytical chemistry at University College, London in 1849 [2, 4]. He became a Fellow of the Royal Society in 1855 [2] and succeeded Thomas Graham (1805–1869) as professor of general chemistry [2, 4], which was combined with his former post. His best and most well known work was on the constitution of alcohol and ether [2–4], especially the Williamson ether synthesis that carries his name [8]. He resigned his position in 1887 and died in the village of Hindhead in Surrey, England in 1904 [2].

[2] Jean Baptiste André Dumas (1800–1884) was born in Alais, in southern France, where he received a classical education before being apprenticed to an apothecary at age 15 [9, 10]. In 1816, he moved to Geneva and entered the pharmaceutical laboratory of Le Royer [9]. He also studied chemistry and attracted the attention of Charles Gaspard de la Rive (1770–1834), who was professor of chemistry at Geneva [9, 10]. During this time, he also met Alexander von Humboldt (1769–1859), who encouraged him to go to Paris to complete his studies. As a result, he became a lecture assistant of Louis Thenard (1777–1857) at the École Polytechnique in 1823 [9, 10]. In 1829, he became one of the founders of the École Centrale des Arts et Manufactures, while also teaching at the École Polytechnique, succeeding Thenard as professor in 1835. He became a member of the Academy of Sciences in 1832 and succeeded Joseph Louis Gay-Lussac (1778–1850) at the Sorbonne as assistant professor, becoming professor in 1841 [9]. Dumas was the first chemist in France to give practical laboratory instruction to students and was considered the most outstanding French chemist of his time [9, 10].

[3] Eugène Melchior Peligot (1811–1890) was an assayer in the Paris Mint, before becoming professor of applied chemistry in the Conservatoire des Arts et Métiers. Besides various collaborations in organic chemistry with Dumas, he discovered potassium chlorochromate (Peligot's salt) and was the first to prepare metallic uranium [11].

1.1 Origin of the Term Alcohol

Fig. 1.1 Organic compounds of the water type

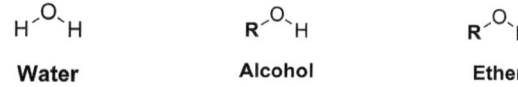

Fig. 1.2 Alexander William Williamson (1824–1904) (Edgar Fahs Smith Collection, University of Pennsylvania Libraries)

[6, 7]. As a result, Jöns Jacob Berzelius (1779–1848)[4] (Fig. 1.5) proposed the general name alcohol for these compounds. Thus ethanol was referred to as wine alcohol (*weinalkohol*) and methanol as wood alcohol (*holzalkohol*) [6]. Shortly thereafter, Dumas and Peligot showed that a compound previously discovered by Michel Chevreul (1746–1889) was cetyl alcohol ($C_{16}H_{33}OH$) [6, 7] and the fact that the family now contained three known examples suggested that a series of such alcohols would be subsequently discovered.

To make its history even more interesting, the origin of the word alcohol does not refer to the substance ethanol, nor does it even derive from any organic species. The word alcohol can ultimately be traced back to the word *kohl* (or *kuhl*),

---

[4] Jöns Jacob Berzelius (1779–1848) was born in a small Swedish town in East Gothland. Both of his parents died when he was young and he was raised by his stepfather Anders Ekmarck. In 1796 he left school after which he entered the University of Uppsala as a medical student. He was forced to leave due to lack of means and became a private tutor. In 1798, however, he won a small scholarship and reentered the University, finally graduating with a dissertation on mineral water [12]. He completed his M.D. in 1802 with a thesis on the medical applications of galvanism and was appointed reader in chemistry at the Carlberg Military Academy in 1806. The following year he was appointed professor of medicine and pharmacy in the School of Surgery in Stockholm, where he had a modest laboratory and funding for apparatus and materials [12]. In 1808 he was elected a member of the Swedish Academy of Sciences and became a joint secretary in 1818 [12]. He resigned his professorship in 1832, but continued to be active in chemical discussions until his death in 1848 [12].

Fig. 1.3 John Dalton's symbol for alcohol from his *New System of Chemical Philosophy* in 1810

Fig. 1.4 Jean Baptiste André Dumas (1800–1884) and Eugène Melchior Peligot (1811–1890) (Edgar Fahs Smith Collection, University of Pennsylvania Libraries)

which referred to a finely powdered form of the mineral stibnite, or antimony trisulphide (Sb_2S_3, Fig. 1.6)[5] [13–17]. Kohl is dark-grey to black and was used as a cosmetic in antiquity, particularly as eye makeup in Egypt. Textual records document its use dating back to at least the 15th century BCE [17].

As Greek and Roman knowledge was eventually transmitted to the Islamic Empire, kohl was modified with the Arabic prefix *al-* to become *al-kohl* (or *al-kuhl*) [13–15, 17, 18]. Over time, the word first came to be used to refer to any very fine powder [13, 15, 17–19] and then further extended to mean the most fine

---

[5] It should be noted that some forms of kohl were comprised of lead sulfide (PbS) [17].

1.1 Origin of the Term Alcohol

Fig. 1.5 Jöns Jacob Berzelius (1779–1848) (Edgar Fahs Smith Collection, University of Pennsylvania Libraries)

Fig. 1.6 Powdered antimony trisulfide (Sb_2S_3), which was known to the ancients as *kohl*

or subtle part of something [14–16]. As a result, *al-kohl* or *al-kohol* was then generally used for any substance attenuated by pulverization, distillation, or sublimation [20]. By the 16th century, Paracelsus[6] (Fig. 1.7) in his *Von Offenen*

---

[6] Paracelsus (1493–1541) was a noted alchemist and medical practitioner whose real name was Philippus Aureolus Theophrastus Bombastus von Hohenheim [21]. Bombastus referred to the

Fig. 1.7 Philippus Aureolus Theophrastus Bombastus von Hohenheim (1493–1541), commonly known as Paracelsus (Edgar Fahs Smith Collection, University of Pennsylvania Libraries)

Schaden and other writings referred to aqueous solutions distilled from wines as *alcool vini* or *alkohol vini* (i.e. the subtle part of wine). Over time, *vini* was then eventually dropped to become first *alkohol* and then finally the modern alcohol [14, 17, 18].

The use of the word alcohol for first antimony and then wine spirits was verified in 1686 by Georg Wedel (1645–1721) in his *Parmacia Acroamatica* [16]:

> …stibio, tanquam collyrio, antimoniato subtilissimo pulvere et alia praeceperunt subtilisari ut alcohol, i.e. ut antimonium pro collyrio tali… de curatione ægritudimun

---

(Footnote 6 continued)
Bombast family of his paternal grandfather and the von Hohenheim did not signify nobility, but the town Hohenheim from which his father came. The nickname Paracelsus was probably given to him by others and adopted by himself because he claimed to be superior to, beyond, or "para" to Celsus, the celebrated 1st century Roman physician [22, 23]. While he learned some medicine and alchemy from his father, it is thought that he entered the University of Basel sometime between 1506 and 1512, but abandoned his studies in 1514 to work in the mines of Tyrol [21, 24]. He studied medicine at Ferrara and claimed to have obtained a M.D. there, but it is debatable whether he actually earned any kind of a medical degree [24]. Never the less, he referred to himself as a doctor of medicine and began practicing medicine with new medicines developed via alchemical methods. In 1526, he was appointed professor of medicine at Basel [23], but his unorthodox methods and outspoken nature ultimately resulted in significant enemies, which forced him to leave his position barely two years later. He spent the last years of his life wandering Germany and Austria, finally dying in 1541 at the age of 48 [21, 25].

oculorum... cohel, vel chofel, vel alchofol, et. significat pulverem debere subtilissimum esse sicut atomi, qui apparent in sphæra solis, intelligitur proprie loquendo per alcohol pulvis subtilisimus. Adæquate verò alcohol etiam prædicatur de subtilitate liquidorum nominatim spirituum, unde spiritus vino alsoholisatus audit, quid adeo subtilis est.

The Latin which can be roughly translated as

...cosmetics, as a salve, powder and other subtle fine antimony specified to alcohol, i.e. antimony as an ointment for the... treatment of the eyes... *cohel* or *chofel* or *alchofol* and the powder should be refined as atoms, which are apparent in the sphere of the sun, it is strictly understood as the fine powder of alcohol. But it is said alcohol adequately specifies too the subtlety of liquid spirits, with wine spirits so heard as alcohol, which is so subtle.

However, even in the 18th century, alcohol was still often defined first as powders of the finest form and only secondly as the spirit of wine [13].

Ethanol as a distinct chemical species, however, was not known until the beginning of the 12th century [14, 15, 18, 19, 26–29]. Prior to that, it comprised an important component of a number of alcoholic beverages produced via fermentation. It is generally believed that the origin of such fermented beverages is shrouded in the mists of human prehistory [22, 30]. While its specific origins are uncertain, it is clear that the production of alcohol via fermentation is one of the oldest forms of chemical technology and the production of beer and wine is thought to predate the smelting of metals.

1.2 Origin of Ethanol Production via Fermentation

The first cases of alcoholic fermentation were most likely a result of serendipity, and quite possibly occurred while early human beings were still nomadic. In addition, the chances of such an event happening only in a single place and time would be expected to be extremely low, and it must be concluded that the consumption of alcohol by fermentation must have been discovered independently by various groups of prehistoric nomadic peoples [30].

Various potential scenarios for the discovery of alcohol by fermentation have been given by authors over the years. Perhaps most common is the natural fermentation of food that had been collected and stored too long. One can easily imagine honey or fruit left to spoil, resulting in the fermentation of such high sugar foods to produce noticeable amounts of alcohol [22, 30]. Others have proposed that this event could have been part of the learning process in the early manipulation and storage of grains during prehistory [31]. Either way, at some point people consumed these fermented materials, most likely driven by thirst, hunger or simple curiosity, and found they liked the taste and the aftereffects.

An alternate theory stems from the observation that under the right conditions, grapes and other berries will literally ferment on the vine. In these cases, the berries are attacked by molds which concentrate the sugar and ultimately yield alcoholic content upon fermentation [22]. Animals and birds will then often eat these berries, especially as food becomes scarcer during the winter months, and early humans could have observed the resulting odd behavior and uncoordinated movements. Again, through either hunger or curiosity, people could have sampled these berries to experience the observed effects firsthand.

The euphoria and mind-altering effects resulting from the consumption of such alcohol-contaminated food must have stimulated early humans to develop methods to produce a regular supply of alcoholic substances from sugar sources available in their local habitats [30, 32]. Because of their intoxicating effects, as well as perceived pharmacological and nutritional benefits, fermented beverages have played key roles in the development of human culture, contributing to the advancement of agriculture, horticulture, and food-processing techniques [32].

Fermented beverages are not thought to predate 10,000 BCE, but were widespread by 4000 BCE [33]. The problem with trying to support the production of fermented beverages during the Paleolithic period (before 10,000 BCE) is the improbability of finding a preserved container with intact organic material or microorganisms that can be identified as exclusively due to such beverages [22, 34, 35]. As such, most authors point to the Neolithic period (ca. 10,000–4000 BCE) as the most probable time for the intentional production of fermented drinks, as this was the first time in human prehistory when the necessary preconditions for this innovation really came together [21, 36–38].

It was during the Neolithic period that early humans transitioned from hunter-gatherers to living in permanent settlements, such as those established on the eastern coast of the Mediterranean from about 10,000 BCE [32]. This area is part of a region referred to as the Fertile Crescent (Fig. 1.8), which stretches from the Nile Delta of northeast Egypt, up the Mediterranean coast to the southeast corner of modern Turkey, and then down again to encompass the land in and around the Tigris and Euphrates rivers [32, 36]. The name stems from a happy accident of geography that provided an ideal environment for wild sheep, goats, cattle, and pigs in the uplands, as well as dense stands of wild grains such as wheat and barley [33]. As such, the area proved ideal for the initial development of permanent, year-round settlements made possible by the availability of animals for hunting and edible plants that could be easily gathered [22, 33, 36].

These initial settlements consisted of simple, round huts, with a typical village consisting of approximately fifty huts, which supported a community of 200–300 people. Although the residents of these villages continued to hunt wild animals, skeletal evidence suggests that they lived on a diet comprised mainly of plants such as acorns, lentils, chickpeas, and various cereals. At this early time period, these were still gathered in the wild, rather than deliberately cultivated [33]. As a result of both a more stable base of operations and a relatively secure food supply, a Neolithic "cuisine" emerged in this region and a variety of food processing techniques were soon developed, including fermentation, soaking, heating, and spicing [22, 34, 36–38]. Thus, the preparation of a range of fermented foods has been known since Neolithic times. The earliest types include leavened bread, beer, and wine via yeasts and cheese via bacteria and molds. These were soon followed

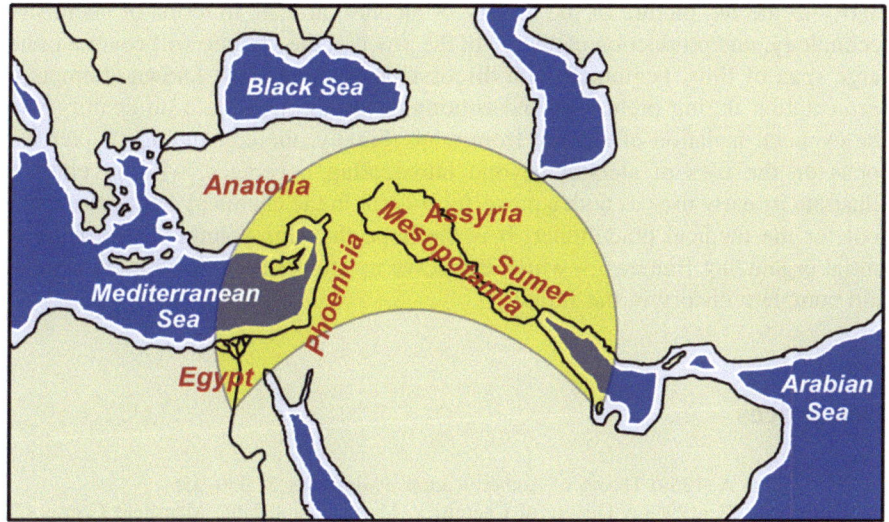

Fig. 1.8 The Fertile Crescent (ca. 4th–2nd millennia BCE)

by East Asian fermented foods such as yogurt and other fermented milk products, as well as a host of traditional alcoholic beverages [22].

The region of the Fertile Crescent is often also called the cradle of civilization, as it saw the development of many of the earliest human civilizations. Thus, it is perhaps not surprising that many of the earliest examples of chemical technology also originate here. In addition to the early production of fermented beverages discussed in the current volume, smelting (Anatolia) and glass production (Syria and Mesopotamia) are both thought to find their origins here [39]. Although the majority of the discussion in this current volume will focus on the developments from the Fertile Crescent, fermented beverages from China were also known during the Neolithic period [32].

1.3 Scope of Current Volume

The history of alcohol and fermented beverages such as beer and wine has been a popular subject for a variety of authors. However, the majority of such publications have focused on either alcohol as a beverage or the social implications of alcoholic drinks. In addition, most reports have focused on one particular type of fermented beverage (i.e. typically either beer or wine) and have thus avoided the complex overlapping histories of the various efforts of mankind to produce and apply fermentation products. Furthermore, little effort has been made to connect the history of fermentation products to the later history of isolated alcohol and other distillation products. This current volume will attempt to provide some clarity to the big picture of the history of alcohol, at least in terms of chemistry, technology, and production methods. In the process, the volume will cover a rather large span of time, beginning with discussions of the earliest known attempts of fermentation during prehistory and continuing up through the 13th century with the eventual isolation of alcohol from wine. Finally, the concluding chapter will focus on the uses of alcohol beyond intoxicating beverages, with an effort to illustrate its early uses as both a powerful solvent in the chemical arts and as a new tool for the medical practitioner. It is the hope that this volume can provide an initial organizing framework which can serve as a foundation for a more detailed and complete history in the future.

References

1. Williamson A (1850) Theory of Aetherification. Philos Mag 37:350–356.
2. Partington, JR (1998) A History of Chemistry, Martino Publishing, Mansfield Centre, CT, Vol. 4, pp 444–454.
3. Harris J, Brock WH (1974) From Giessen to Gower Street: Towards a Biography of Alexander William Williamson (1824–1904). Ann Sci 31:95–130.
4. Inde AJ (1964) The Development of Modern Chemistry. Harper & Row, New York, pp 212–215.
5. Crosland MP (1978) Historical Studies in the Language of Chemistry. Dover Publications, Inc., New York, pp 260–261.
6. Partington, JR (1998) A History of Chemistry, Martino Publishing, Mansfield Centre, CT, Vol. 4, p 353.
7. Inde AJ (1964) The Development of Modern Chemistry. Harper & Row, New York, p 189.
8. Vollhardt KPC (1987) Organic Chemistry. W. H. Freeman and Company, New York, pp 327–328.
9. Partington, JR (1998) A History of Chemistry, Martino Publishing, Mansfield Centre, CT, Vol. 4, pp 337–340.
10. Inde AJ (1964) The Development of Modern Chemistry. Harper & Row, New York, pp 150–153.
11. Partington, JR (1998) A History of Chemistry, Martino Publishing, Mansfield Centre, CT, Vol. 4, p 362.
12. Partington, JR (1998) A History of Chemistry, Martino Publishing, Mansfield Centre, CT, Vol. 4, pp 142–144.
13. Crosland MP (1978) Historical Studies in the Language of Chemistry. Dover Publications, Inc., New York, pp 107–108.
14. Stillman JM (1924) The Story of Early Chemistry. D. Appleton and Co., New York, pp 187–192.
15. Forbes, RJ (1970) A Short History of the Art of Distillation. E. J. Brill, Leiden, pp 87–90.
16. Partington, JR (1998) A History of Chemistry, Martino Publishing, Mansfield Centre, CT, Vol. 2, p 316.
17. von Lippmann EO (1912) Zur Geschichte des Alkohols und seines Namens. Angew Chem 40:2061–2065.
18. Leicester HM (1971) The Historical Background of Chemistry. Dover Publications, Inc., New York, pp 76–77.
19. Liebmann AJ (1956) History of Distillation. J Chem Educ 33:166–173.
20. Forbes, RJ (1970) A Short History of the Art of Distillation. E. J. Brill, Leiden, pp 47–107
21. Holmyard, EJ (1990) Alchemy. Dover, New York, pp 165–169.

References

22. McGovern PE, Hartung U, Badler VR, Glusker DL, Exner LJ (1997) The beginnings of winemaking and viniculture in the ancient Near East and Egypt. Expedition 39:3–21.
23. Read, J (1995) From Alchemy to Chemistry. Dover, New York, pp 97–98.
24. Titley, AF (1938) Paracelsus. A résumé of some controversies. Ambix 1:166–183.
25. Stillman JM (1924) The Story of Early Chemistry. D. Appleton and Co., New York, p 310.
26. von Lippmann EO (1920) Zur Geschichte des Alkohols. Chem Ztg 44:625.
27. Forbes RJ (1970) A Short History of the Art of Distillation. E. J. Brill, Leiden, pp 55–65.
28. Gwei-Djen L, Needham J, Needham D (1972) The Coming of Ardent Water. Ambix 19:69–112.
29. Vallee BL (1998) Alcohol in the Western World. Sci Am 279(June):80–85.
30. Hornsey IS (2003) A History of Beer and Brewing. The Royal Society of Chemistry, Cambrige, pp 1–3.
31. Joffe AH (1998) Alcohol and Social Complexity in Ancient Western Asia. Curr Anthropol 39:297–322.
32. McGovern PE, Zhang J, Tang J, Zhang Z, Hall GR, Moreau RA, Nunez A, Butrym ED, Richards MP, Wang C, Cheng G, Zhao Z, Wang C (2004) Fermented beverages of pre- and proto-historic China. Proc Natl Acad Sci USA 101:17593–17598.
33. Standage, T (2005) A History of the World in 6 Glasses. Walker & Company, New York, pp 10–15.
34. McGovern PE (2012) The Archaeological and Chemical Hunt for the Origins of Viniculture in the Near East and Etruria. In Archeologia della Vite e del Vino in Toscano e nel Lazio: Dalle tecniche dell'indagine archeologica alle prospettive della biologia molecolare. Ciacci A, Rendini P, Zifferero A (eds). Edizioni all'Insegna del Giglio: Borgo S. Lorenzo, pp 141–152.
35. McGovern PE, Mirzoian A, Hall GR (2009) Proc Natl Acad Sci USA 106:7361–7366.
36. Cavalieri D, McGovern PE, Hartl DL, Mortimer R, Polsinelli M (2003) Evidence for *S. cerevisiae* Fermentation in Ancient Wine. J Mol Evol 57:S226–S232.
37. McGovern P (2002) Wine and the Vine: New Archaeological and Chemical Perspectives on Its Earliest History. In Bacchus to the Future. C. Cullen C, Pickering G, Phillips R (eds) Brock University: St. Catherines (Ontario, Canada), pp 565–592.
38. McGovern PE (2011) The Archaeological and Chemical Hunt for the Origins of Viniculture. In Perard P, Perrot M (eds) Recontres du Clos-Vougeot 2010. Centre Georges Chevrier, Dijon, 13–23.
39. Rasmussen SC (2012) How Glass changed the World. The History and Chemistry of Glass from Antiquity to the 13th Century. Springer Briefs in Molecular Science: History of Chemistry. Springer, Heidelberg.

Chapter 2
Earliest Fermented Beverages

Fermented beverages may be made from a variety of sugar-containing materials: cider from apples, sake from rice, mead from honey, beer from grains, and of course, wine from grapes [1]. However, if one considers the primary sources of sugar available to early peoples, this quickly becomes limited to wild berries and other fruits, tree sap, honey, and possibly milk from animals. In warmer climates, these sources could have been relatively plentiful, even in pre-farming eras, but in temperate zones, there would have been few abundant sources of sugar other than honey [2]. As a result, the earliest alcoholic beverage is generally thought to have been fermented honey [2, 3].

2.1 Fermentation

Alcoholic fermentation, also referred to as ethanol fermentation, is a biological process in which yeasts obtain energy via the conversion of various sugars (Fig. 2.1) into ethanol and carbon dioxide. Yeasts are eukaryotic microorganisms classified in the kingdom Fungi and are estimated to be approximately one percent of all fungal species. The principal yeast species responsible for fermentation is *Saccharomyces cerevisiae*, which has been used for thousands of years in both baking and the production of alcoholic beverages [4–6]. Evidence for this was found in material collected from Egyptian wine jars dating back to ~3150 BCE. This material, which was presumed to include dead cells and cellular debris from yeast, was extracted and subjected to two independent sets of PCR amplifications with the appropriate negative and positive controls. The resulting analysis then revealed DNA from an organism that can confidently be assigned to *Saccharomyces cerevisiae* [4]. Of course, there exists a wide variety of *S. cerevisiae* strains [6], with differences in kinds of yeast recognized as early as 1550 BCE, as demonstrated by the various yeasts types mentioned in the Ebers Papyrus: wine yeast, beer yeast and mesta-yeast, growing yeast, bottom yeast, yeast juice, and yeast water [7]. In addition, it has been postulated that early craftsmen cultivated successful yeast strains, possibly skimming off the frothy yeast from the surface of the fermentation vat in order to use it again for later fermentations [4]. It has even been suggested that yeast was possibly cultivated before grain [1] and may be the oldest domesticated plant [8].

The study of alcoholic fermentation dates back to efforts of Antoine Lavoisier[1] (1743–1794) (Fig. 2.2), who described the phenomenon as "one of the most extraordinary in chemistry" [9]. As a result, Lavoisier published in 1789 the first clear account of the chemical changes that occur in fermentation, reporting that 100 parts by weight of sugar were converted to 60.17 parts of alcohol, 36.81 parts of carbon dioxide and 2.61 parts of acetic acid. After developing improved analytical methods with Louis Thenard (1777–1857), Joseph Louis Gay-Lussac[2] (1778–1850) (Fig. 2.2) later revised Lavoisier's figures in 1810, estimating that fermentation of 100 parts of sugar resulted in 51.34 parts of alcohol and 48.66 parts of carbon dioxide [9]. Perhaps as a result of this work, the overall modern equation for alcoholic fermentation

$$C_6H_{12}O_6 \rightarrow 2\ C_2H_5OH + 2\ CO_2$$

is commonly attributed to Gay-Lussac, with most citing his work of 1815. However, as clearly pointed out by Barrett [8], the empirical formula for glucose was not established until 1843 by Dumas, with the molecular formula finally being published in 1870 by Adolf von Baeyer (1835–1917). Thus, as Gay-Lussac died in 1850, the modern equation given above cannot be correctly credited to him.

In reality, the simplistic equation above is somewhat deceptive as it ignores the fact that alcoholic fermentation is catalyzed by a number of enzymes and cofactors

---

[1] Antoine Laurent Lavoisier (1743–1794) was born in Paris on August 26, 1743 to a family of commoners that was just beginning to attain some status [10, 11]. He was educated at the College Mazarin, where he soon developed a taste for mathematics and physical science [11, 12]. He pursued legal studies as his main interest, however, eventually receiving a Bachelors of Law in 1763 [10, 11]. He eventually became a businessman, astronomer, geologist, tax collector, and a noted chemist. He was a member of the French financial and government establishment and was at one time or another President of the Academy of Sciences, Chief of the Bureau of Accounts, Commissioner to the National Treasury, and member of the National Assembly [10–13]. He is considered by many as the father of modern chemistry and included among his greatest accomplishments were a new, logical system of nomenclature and the organization of a new system of chemistry, with an operational definition of elements, classification of reactions and composition, and the mass balanced equation [11–13]. During the Reign of Terror, he was tried and sentenced as a result of his positions within the old regime, leading to his death by guillotine on May 8th, 1794 [10–13].

[2] Joseph Louis Gay-Lussac (1778–1850) entered the École Polytechnique in 1797. There he attracted the notice of the professor of chemistry, Claude Louis Berthollet (1748–1822), and began working in his private laboratory [14–16]. He received his Master of Arts in 1800 and went on to act as a demonstrator to Antoine François de Fourcroy (1755–1809). In 1809, he became a professor of chemistry in the École Polytechnique, as well as a professor of physics in the Sorbonne [15, 16]. He resigned from the Sorbonne in 1832 and became professor of chemistry at the Jardin des Plantes [15, 16]. Equally proficient both in chemistry and physics, Gay-Lussac made his mark in both sciences and is probably best known for his quantitative studies of the properties of gases [14, 16]. Gay-Lussac died in Paris in 1850.

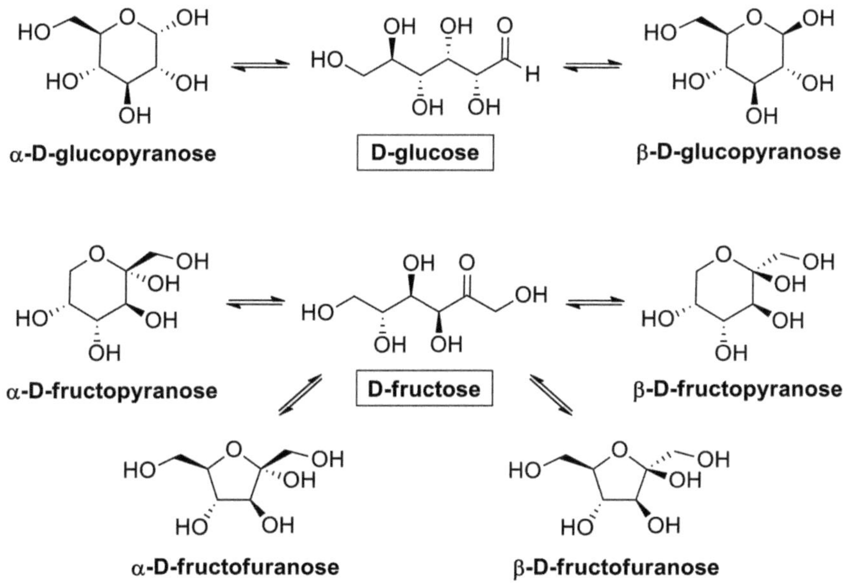

2.1 Fermentation

Fig. 2.1 The sugars D-glucose and D-fructose in the linear form, as well as the various cyclic hemiacetal isomers that predominate in solution

Fig. 2.2 Antoine Lavoisier (1743–1794) and Joseph Louis Gay-Lussac (1778–1850). (Edgar Fahs Smith Collection, University of Pennsylvania Libraries)

Fig. 2.3 Eduard Buchner (1860–1917). (Edgar Fahs Smith Collection, University of Pennsylvania Libraries)

that occur naturally in yeast [1, 17]. This mixture was extracted from yeast by Eduard Buchner[3] (1860–1917) (Fig. 2.3) in 1897, which allowed the first alcoholic fermentation outside of the cell. Buchner thought the extract contained an enzyme that catalyzed the fermentation [17, 19–21], stating [22]:

> The active agent in the expressed yeast juice appears rather to be a chemical substance, an enzyme, which I have called "zymase".

Thus, a more complete form of the equation above would give

$$C_6H_{12}O_6 \xrightarrow{zymase} 2\ C_2H_5OH + 2\ CO_2$$

---

[3] Eduard Buchner (1860–1917) was born in Munich on May 20, 1860 [18], the son of Ernst Buchner, professor of forensic medicine and physician at the University of Munich [19]. After a short period of study at the Munich Polytechnic, where he attended courses by Emil Erlenmeyer (1835–1909), he worked in a preserve and canning factory for four years to raise money for further studies [19]. He returned to his studies in 1884, including organic chemistry under Adolf von Baeyer [18] and plant physiology under Carl Wilhelm von Nägeli (1817–1891) [18, 19], and taking his doctor's degree at the University of Munich in 1888 [18]. von Baeyer took Buchner on as a paid assistant in 1890, made possible his Habilitation in 1891, and arranged for generous funding from the Munich brewers, which made it possible for Buchner to establish a small laboratory for the chemistry of fermentation [19]. In 1894 Buchner took an interim appointment at the University of Kiel [18, 19] before being granted the title of Professor in 1895. In 1896 he was appointed professor for analytical and pharmaceutical chemistry at the University of Tübingen, before becoming the chair of general chemistry in the Agricultural College in Berlin and director of the Institute for the Fermentation Industry in 1898 [18, 20]. He was awarded the Nobel Prize in 1907 for his biochemical investigations and his discovery of non-cellular fermentation. In 1909 Buchner moved to Breslau to the chair of physiological chemistry and in 1911 to Wurzburg [18, 20]. He died in 1917 as a result of wounds received on the Romanian front during the First World War [18].

2.1 Fermentation

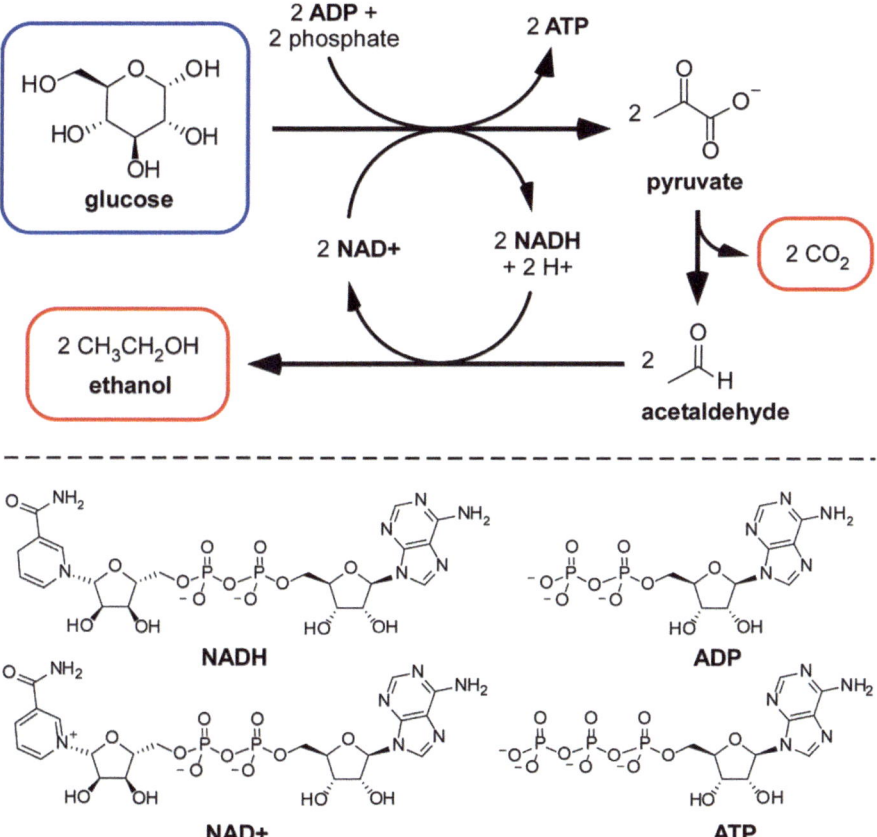

Fig. 2.4 A simplified outline of the alcoholic fermentation process

It was only later that it was determined that *zymase* was a complex mixture of enzymes and not a single species [1]. A simplified scheme of the alcoholic fermentation process is outlined in Fig. 2.4, in which the first overall step shown (i.e. glucose to pyruvate) actually represents ten individual steps requiring ten different enzymes [17, 23]. In addition to the major products of ethanol and carbon dioxide, there are a large number of other minor by-products of yeast metabolism that contribute to the flavor of different fermented products [5]. In fact, the theoretical conversion of 180 g sugar into 92 g ethanol (51.1 % by weight) and 88 g carbon dioxide (48.9 % by weight) could only be expected in the absence of any yeast growth, production of other metabolites, or loss of ethanol as vapor [23]. In a model fermentation process, about 95 % of the sugar is converted into ethanol and carbon dioxide, with 1 % converted into cellular material, and the remaining 4 % into other products such as glycerol [23].

In the production of fermented beverages, ethanol is the primary product with carbon dioxide as a significant by-product. In the case of beer, the dissolved gas gives its carbonation and creates the head or froth on the beer. This additionally assists in preservation as a sealed cask that is retained under the pressure of carbon dioxide keeps air and microbes out of the empty portion of the cask [1].

This same fermentation process is utilized in the production of leavened breads. In this case, carbon dioxide is caught in the glutinous material, causing it to rise and producing bread that is soft, light, and fluffy in texture. Ethanol is still produced in this process, but is lost at the elevated temperatures used during baking [1].

2.2 Mead

The collection of bee honey is an ancient activity and is thought to date back into the Paleolithic period (before 10,000 BCE) [24]. Evidence of this has been found in Mesolithic rock art from Cuevas de la Araña near Valencia, Spain that dates to 6000–8000 BCE. As shown in Fig. 2.5, this rock art illustrates honey collection from a wild nest. A number of such rock art examples have been found in Europe, Africa, Asia, and Australia [24].

As such, honey provided an abundant source of sugar, from which an alcoholic beverage could be produced by the fermentation of aqueous honey solutions. This drink is referred to as mead,[4] hydromel, or honey wine (Table 2.1), and is known from many sources of ancient history throughout Europe, Africa, and Asia [2, 25]. It is thought that such fermented honey drinks were the first intoxicating beverage made by most primitive peoples and were in fact made thousands of years before either wine or beer were produced [2, 25, 27–29]. Archaeological evidence of fermented honey, however, is somewhat ambiguous. The problem lies in that the archeological confirmation of either beeswax or certain types of pollen is indicative of the presence of honey, but not necessarily its fermentation. As honey has been used historically to sweeten various drinks, the detection of honey alone cannot be correlated to the production of mead [2].

One of the earliest items of material evidence of mead is a drinking horn recovered at Skudstrup, in northern Germany. The horn was buried in a peat bog and has been dated to before 100 CE. Analysis of the horn gave evidence of both yeasts and pollen grains, suggesting that it had held mead [27, 29, 30].

More recent archaeological evidence also suggests the use of honey in China for the early production of fermented drinks based upon the chemical analysis of pottery sherds dating to 7000–5500 BCE from Jiahu, an early Neolithic village in the Henan province of China [31]. Material extracted from the sherds were analyzed by Fourier-transform infrared spectroscopy (FTIR), gas chromatography–mass spectrometry (GC–MS), and high-performance liquid chromatography–mass

---

[4] The English word *mead* derives from the Old English *meodu*, a derivative of the Proto-Indo-European root *médhu* (honey, or an alcoholic drink fermented from honey). While in the east, *médhu* referred to both honey and its fermented drink, its use in the west referred only to the fermented drink and *melit* was used to denote honey [26].

Fig. 2.5 The reproduction of a portion of Mesolithic rock art from Cuevas de la Araña near Valencia, Spain (created by Achillea, 2007)

Table 2.1 Fermented beverages made with honey [27]

Beverage name	Description
Mead (or hydromel)	Fermented mixture of honey and water
Sack mead	Fermented mixture of honey with less water
Metheglin	Fermented mixture of honey, water, and spices
Mulsum	Wine mixed with honey
Pyment (or clarree)	Wine mixed with honey and spices

spectrometry (HPLC–MS), which revealed the presence of long straight-chain hydrocarbons (C_{23}–C_{36}) attributed to the presence of beeswax and IR signals attributed to tartaric acid (a common biomarker for grapes and other select fruits). The presence of tartaric acid was further supported by positive Feigl spot tests [31]. Based upon these analyses and spectral matching of the FTIR and HPLC analyses with various modern samples, the authors suggest that the pottery contained a mixed fermented beverage containing honey, rice, and a fruit, the latter postulated to be Chinese hawthorn.

Further archaeological evidence suggesting the presence of mead was reported as a result of the analysis of various bronze drinking vessels from the tomb of King Midas (ca. 700 BCE) in central Turkey [32]. As with the previous analysis in China, a combination of FTIR, liquid and gas chromatography, and mass spectrometry were used to detect tartaric acid and its salts as biomarkers of wine, calcium oxalate ("beerstone") as a biomarker of beer, and hydrocarbons indicative of beeswax. This combination led the authors to postulate that the vessels contained a mixture of grape wine, beer, and mead [32]. The reliability of calcium oxalate and tartaric acid as biomarkers for beer and wine will be addressed in detail in Chaps. 3 and 4, respectively.

However, as discussed above, all of these analyses suggest the presence of honey, but provide no direct evidence of its fermentation. The correlation of the detection of honey to it being a component of mead is usually based upon the container (i.e. drinking horn, drinking vessels) and the environment in which the material under analysis was found, as well as simple postulation by those carrying out the analyses.

The earliest written evidence of mead is believed to come from the *Rigveda*, an Indian collection of Vedic Sanskrit hymns, dated to around 1700–1100 BCE. This book contains approximately 300 references to *mádhu*,[5] the Sanskrit word for both honey and mead [26–28]. It is unclear, however, if any of these occurrences of *mádhu* actually referred to mead [27, 28]. Another Indian text dated to the 4th–5th century BCE, the *Ramayana*, mentions becoming intoxicated after drinking *mádhu*, likely referring to mead in this case [27]. Mead (as hydromel) [33] was specifically discussed by the first century historian Pliny the Elder[6] in his *Naturalis Historia* (*Natural History*) and differentiates it from wine sweetened with honey (a drink known as mulsum, Table 2.1) [34]. Lucius Junius Moderatus Columella,[7] considered one of the most important writers on agriculture of the Roman empire, also discusses mead in his *De Re Rustica* (written in 42 CE) [27, 35].

---

[5] The Sanskrit word *mádhu* has been defined as: honey; the juice or nectar of flowers; anything sweet; mead, sugar, liquorice; sweetness; a spirituous liquor obtained from the blossoms of the *Bassia latifolia*; any sweet intoxicating drink; wine; spirituous liquor [26].

[6] Pliny the Elder or Gaius Plinius Secundus (23–79 CE) was a Roman officer and encyclopedist. Born in late 23 or early 24 at Novum Comum (modern Como), a small city in the region known as Transpadane Gaul (or Gallia Transpadana), he was introduced to Rome at an early age. He studied in Rome before becoming a military tribune at age 21. During his time as an officer, he held three posts, serving primarily in Germany. Pliny is best known as a writer and encyclopedist, writing his first treatise in 50–51, followed by a two-volume biography of the senator Pomponius Secundus, and the twenty-volume History of Rome's German Wars. He is most well-known for his encyclopedia, *Naturalis Historia*, published in 77 CE. This massive work resulted from years of collecting records, both from his own reading and from personal observations, as well as anything else that seemed to him worth knowing. He died in late August of 79 during the evacuation around the erupting volcano Vesuvius. The exact cause of his death is unknown, but it has been said that he was asthmatic and overcome by sulfurous fumes. Reports are that he was still recording his personal observations of the event during the final hours of his life [37].

[7] Lucius Junius Moderatus Columella (4 CE–ca. 70 CE) is considered the most important writer on agriculture of the Roman Empire. He came from provincial Spain, but moved to Italy as a young man, where he took up farming and lived near Rome [38]. He is most well-known for his *De Re Rustica*, which comprised twelve volumes on farming, animal husbandry, and estate management, and this work forms an important source on Roman agriculture.

2.2 Mead

The ferment for mead is the osmophilic yeasts[8] naturally present in honey [27, 29]. Most of the known osmophilic yeasts are included in the species *Saccharomyces rouxii*, *Saccharomyces bailii* var. *osmophilus*, and *Saccharomyces bisporus* var. *mellis* [36]. As such yeasts are already present in the honey, its fermentation can occur quite easily without effort. For example, unintentional fermentation could occur if rain simply fell into a vessel containing honey combs or honey [27]. The ease of its production is illustrated by the following preparation of mead as described by Pliny the Elder in his *Naturalis Historia* [33]:

> There is a wine also made solely of honey and water. For this purpose it is recommended that rain-water should be kept for a period of five years. Those who show greater skill, content themselves with taking the water just after it has fallen, and boiling it down to one third, to which they then add one third in quantity of old honey, and keep the mixture exposed to the rays of a hot sun for forty days after the rising of the Dog-star; others, however, rack it off in the course of ten days, and tightly cork the vessels in which it is kept. This beverage is known as "hydromeli," and with age acquires the flavour of wine. It is nowhere more highly esteemed than in Phrygia.

An earlier and more detailed preparation of mead is also given by Columella in his *De re rustica* [35]:

> Therefore having set apart this bees-wax-water, and destinated it for preserving of fruits, mead must be made by itself of the very best honey; but it is not made after one manner: for some, many years before, put up rain-water in vessels, and set it in the Sun in the open air; then, having emptied it from one vessel to another, and made it very clear (for, as often as it is poured from one vessel to another, even for a long time, there is found, in the bottom of the vessel, some thick settling like dregs) they mix a sextarius[9] of old water with a pound of honey.
>
> Nevertheless some, when they have a mind to make the mead of a rougher taste, mingle a sextarius of water with three quarters of a pound of honey; and after they have, according to this proportion, filled a stone bottle, and plaistered it, they suffer it to be forty days in the Sun, during the rising of the Dog-star; then they put it up in a lost, which receives smoak. Some, who have not been at the pains to preserve rain-water till it becomes old, take that which is fresh, and boil it into a fourth part: then, after it is grown cold, if either they have a mind to make mead sweeter than ordinary, they mix a sextarius of honey with two sextarii of water; or, if they would have it rougher, they put three quarters of a pound of honey to a sextarius of water; and, having made it according to these proportions, they pour it into a stone bottle; and, after they have kept it forty days in the Sun, as I said above, they put it up in a lost, which receives smoak from below.

Honey wines can be made with alcohol content of up to 10–12 % if fermented long enough, and are often sweet due to significant amounts of unfermented sugars [29, 39, 40]. The osmophilic yeasts present in honey are best for the fermentation of honey solutions with sugar concentrations above 15 %, but generally do not produce alcohol as well as the common yeasts of beer and wine, *Saccharomyces cerevisiae* [29]. *Saccharomyces cerevisiae* species are the best for fermentation of

---

[8] The term osmophilic yeast, coined by A. A. von Richter in 1912, has been used to designate yeast strains that are able to thrive in highly concentrated sugar solutions [36].

[9] The theoretical value for the sextarius is about 540.3 ml.

honey, providing the sugar concentration is less than 15 %. However, through the use of various additives, the fermentation of honey by *Saccharomyces cerevisiae* at sugar concentrations of 25 % have been successfully reported, resulting in alcohol concentrations of 12–15 % [29, 39].

Mead can be made from nearly any type of honey and the resulting mead produced retains many of the characteristics of the honey utilized. For example, light honeys yield lighter meads, while dark honeys of stronger flavor are typically favored for the making of honey ales or a more beer-like drink [29, 39]. Mead can also be produced by fermenting an aqueous solution of honey with grain mash, which is then strained after fermentation. Meads can also be flavored with spices (metheglin, Table 2.1), fruit, or hops, the latter of which produces a bitter, beer-like flavor.

As wine was produced in the Mediterranean region and the warm temperate parts of Europe, it obtained a higher social status and displaced honey-based alcoholic drinks. Elsewhere, however, beverages made by fermenting honey often remained important [27]. This was particularly true in the Germanic north [26] and mead was popular in central and northern Europe at least as early as 334 BCE [29].

2.3 Date Wine

Perhaps the most abundant source of sugar in the Fertile Crescent came from the date palm (*Phoenix dactylifera* L., Fig. 2.6), and offers the most likely means by which alcoholic drinks were first produced with any degree of regularity [2, 41]. The date palm was characteristic of the whole of Babylonia from the oldest period, and while it was not indigenous to North Mesopotamia, Syria, or Asia Minor [41], it was still prolific in both Mesopotamia and Egypt [42]. By 4000 BCE, the date palm had been domesticated, most probably in southern Mesopotamia [2, 43, 44] where it was well established by 3000 BCE [45]. The palm is adapted to a xeric habitat and thrives well in desert environments unsuitable for most crops [45]. While it thrives in a hot and dry climate, it does require plenty of water. The date palm, however, can tolerate water of high salinity, much more so than any other cultivated tree [43, 44]. In Palestine, its natural habitat is the Rift Valley, though some small plantations might also be found in the southern coastal plain [43].

The tree furnishes both fruit (dates) and sap. The fruit itself is rich in sugar (dry dates contain 70–80 % sugar [2, 43–45]) and provided a source of honey (i.e. a date-sugar syrup) [41]. As the fruit itself contains the necessary yeast, fermentation of the fruit or honey is fairly rapid in these warm climates [42], thus facilitating the production of both wine and vinegar [41]. Economically, the date palm is a very worthwhile plant with significantly high productivity [43] and the ancient Egyptians considered it the most important of the fruit trees cultivated [46]. Trees can bear fruit for 60–100 years and the average palm tree can produce as much as 40 kg of fruit per year [44], while very productive trees can produce as much as 100 kg or more each year [45]. In addition, it has been reported that in modern

2.3 Date Wine

Fig. 2.6 *Phoenix dactylifera* L. (Botanical print (dated 1884); photo by Symac 2004)

Israeli plantations, the annual harvest of a typical tree is 100–200 kg of sugar. As such, up to 10 tons of sugar can be achieved with a relatively small investment of labor [43].

As with other fermented beverages, it is unknown exactly when the production of date wine (or date beer, as it is sometimes referred) began [43]. The oldest solid evidence of date wine comes from the period of 3000–2000 BCE [42, 44]. However, it is believed to have been prepared by the people in Mesopotamia and Egypt long before that and it is thought that date wine probably preceded the production of barley beer [42, 43]. It is believed by some that the date palm and barley provide the first direct evidence that both the Mesopotamians and the Egyptians were making fermented drinks [42]. In Egypt, however, the consumption of date wine (*bená*) was limited to the lower classes [7]. In a similar manner, it was largely secondary to barley beer in Mesopotamia until the beginning of the Iron Age, after which it then became the principal alcoholic beverage of choice [43]. Pliny the Elder described date wine as one of the varieties of artificial wines (i.e. non-grape wines) and stated that it was prepared by the Parthians[10] and the Indians, as well as throughout all the countries of the East [47].

---

[10] Parthia was a region of north-eastern Persia.

The production of date wine was not a very complex process and was much simpler than that required for cereal beers [42, 43]. Fundamentally one only required a container in which to put the date mash during fermentation and some device with which to strain it when complete [42]. However, in order to produce the mash, the fruit first required pressing [43, 44]. It is theorized that the dates were initially pitted, as crushed pits in the mash might be a nuisance. In addition, the date pits are known to have nutritional value, similar to barley. In modern times, shredded or ground date pits are served as animal fodder and in times of desperation can be added to wheat or barley flour [43].

Pressing the fruit could be accomplished by simple foot-treading, but it is much more efficient to press it with heavy stone rollers. Four such rollers have been found so far, with one large roller thought to weigh 900 kg discovered at Ein Feshkha, on the northwestern shore of the Dead Sea [43]. It is proposed that such rollers must have been operated by two or three people. Rolling such a heavy cylindrical stone, however, was thought to have been not that difficult as task, as the crushed dates could theoretically serve as a lubricant [43].

A nearly equal amount of water was then added to the mash [42]. In his *Naturalis Historia*, Pliny the Elder described this as follows [47]:

A modius[11] of the kind of ripe date called "chydææ" is added to three congii[12] of water, and after being steeped for some time, they are subjected to pressure.

As described by Pliny, this ratio would be approximately 0.83 equivalents of dates to each equivalent of water. In Egypt, the date wine was made in January when the water used from the Nile was the clearest [7]. This mixture was then allowed to undergo fermentation and strained. The same dates could be used a number of times, as the yeast stops functioning when the alcohol content becomes too high, thus ending the fermentation. After straining, fresh water could be added to the remaining date mash and the fermentation processes could be repeated again [43]. Date wines were also sometimes flavored with herbs or other additives. Numerous varieties of date wines have been described in surviving records and it has been proposed that the additions made to these drinks and the exact methods of manufacture were jealously guarded secrets [42].

2.4 Palm Wine

Palm wine is also produced from the date palm, but differs in that it is the sap that is collected and fermented, rather than the fruit [2, 7, 44]. The sap is said to consist of about 10 % sugar and significant amounts of sap can be extracted from the palm

---

[11] The modius was equal to 16 sextarii or a little more than the peck, a unit of dry volume equivalent to 2 gallons.

[12] The congius was a liquid measure equivalent to approximately 3.48 L or 0.92 U.S. gallons.

tree. Quantities quoted per tree have been given as high as 8 L a day and up to 400–500 L a year. However, a tree used for its sap is significantly weakened [44], causing it to stop bearing fruit [43, 46, 48]. It has been reported that it then requires at least 4–5 years to recover enough to do so again [40], although others have noted that the tree often dies as a consequence [46, 48]. As such, sap is generally harvested only from male or fruitless old trees [43, 44], or in regions in which they grow in great abundance [46].

Palm wine was produced by first gathering the sap, which was accomplished by making an incision in the heart of the date tree below the base of the upper branches [46, 48]. A jar was then attached to this part of the tree in order to catch the juice which flowed from the incision [46]. The liquid extracted from date trees can be drunk in large quantity, but after fermentation it became a powerfully intoxicating beverage [46, 48]. Fermentation occurs rapidly and sap collected in the morning can contain 4–5 % alcohol by the same evening.

In Egypt, palm wine (*áama*) was drunk by the lower classes in the same way as date wine [7]. In addition to its production in Egypt, it is highly probable that such wine was also produced in ancient Palestine, although there is not explicit evidence of such. However, 'palm water' is mentioned in the Mishnah[13] as a thirst quencher, and it is thought that such 'palm water' would have been fermented for wine [43].

Palm wine is still being produced today, particularly in Egypt (where it is now called *lowbgeh*) and Libya (where it is called *lakbi*) [7, 43, 46, 48]. It is also much used in West Africa and Madagascar, as well as most of the East Indian islands, where it is known as Indian toddy [7].

References

1. Lambert JB (1997) Traces of the Past. Unraveling the Secrets of Archaeology through Chemistry. Addison-Wesley, Reading, MA, pp 134–136.
2. Hornsey IS (2003) A History of Beer and Brewing. The Royal Society of Chemistry, Cambridge, pp 6–7.
3. Vallee BL (1998) Alcohol in the Western World. Sci Am June 1:80–85.
4. Cavalieri D, McGovern PE, Hartl DL, Mortimer R, Polsinelli M (2003) Evidence for *S. cerevisiae* Fermentation in Ancient Wine. J Mol Evol 57:S226–S232.
5. Donalies UEB, Nguyen HTT, Stahl U, Nevoigt E (2008) Improvement of *Saccharomyces* Yeast Strains Used in Brewing, Wine Making and Baking. Adv Biochem Engin Biotechnol 111:67–98.
6. Legras JL, Merdinoglu D, Cornuet JM, Karst F (2007) Bread, beer and wine: Saccharomyces cerevisiae diversity reflect human history. Mol Ecol 16:2091–2102.
7. Partington, JR (1935) Origins and Development of Applied Chemistry. Longmans, Green and Co., London, pp 197–198.

---

[13] The Mishnah is a Jewish text which represents the first major written record of the Jewish oral traditions.

8. Braidwood RJ (1953) Symposium: Did man once live by beer alone? Am Anthro 55:515–526.
9. Barnett JA (1998) A History of Research on Yeasts 1: Work by Chemists and Biologists 1789–1850. Yeast 14:1439–1451.
10. Partington, JR (1998) A History of Chemistry, Martino Publishing, Mansfield Centre, CT, Vol. 3, pp 363–368.
11. Stillman JM (1924) The Story of Early Chemistry. D. Appleton and Co., New York, pp 513–539.
12. Bell MS (2005) Lavoisier in the Year One. W. W. Norton & Company, New York.
13. Inde AJ (1964) The Development of Modern Chemistry. Harper & Row, New York, pp 58–86.
14. Inde AJ (1964) The Development of Modern Chemistry. Harper & Row, New York, pp 116–118.
15. Partington, JR (1998) A History of Chemistry, Martino Publishing, Mansfield Centre, CT, Vol. 4, pp 77–90.
16. Partington JR (1950) J. L. Gay-Lussac (1778–1850). Nature 165:708–709.
17. Barnett JA (2002) A history of research on yeasts 5: the fermentation pathway. Yeast 20:509–543.
18. Manchester K (2000) Biochemistry comes of age: a century of endeavour. Endeavour 24:22–27.
19. Lothar Jaenicke L (2007) Centenary of the Award of a Nobel Prize to Eduard Buchner, the Father of Biochemistry in a Test Tube and Thus of Experimental Molecular Bioscience. Angew Chem Int Ed 46:6776–6782.
20. Barnett JA, Lichtenthaler FW (2001) A history of research on yeasts 3: Emil Fischer, Eduard Buchner and their contemporaries, 1880-1900. Yeast 18:363–388.
21. Kohler R (1971) The background to Eduard Buchner's discovery of cell-free fermentation. J Hist Bio 4:35–61.
22. Buchner E (1966) Cell-free fermentation. Nobel Lecture, December 11, 1907. In: Nobel Lectures, Chemistry 1901–1921. Elsevier Publishing Company, Amsterdam.
23. Pretorius IS (2000) Tailoring wine yeast for the new millennium: novel approaches to the ancient art of winemaking. Yeast 16:675–729.
24. Crane E (1999) The World History of Beekeeping and Honey Hunting. Routledge, New York, pp 36–42.
25. Arnold JP (1911) Origin and History of Beer and Brewing, From Prehistoric Times to the Beginning of Brewing Science and Technology. Alumni Association of the Wahl-Henius Institute of Fermentology, Chicago, p 149.
26. Le Sage DE (1975) The Language of Honey. In: Crane E (ed) Honey. Crane, Russak & Company, Inc., New York.
27. Crane E (1999) The World History of Beekeeping and Honey Hunting. Routledge, New York, pp. 513–515.
28. Arnold JP (1911) Origin and History of Beer and Brewing, From Prehistoric Times to the Beginning of Brewing Science and Technology. Alumni Association of the Wahl-Henius Institute of Fermentology, Chicago, p 44.
29. Morse RA, Steinkraus KH, Paterson PD (1975) Wines from the Fermentation of Honey. In: Crane E (ed) Honey. Crane, Russak & Company, Inc., New York.
30. Betts AD (1932) Nectar Yeasts. Bee World 13:115–116.
31. McGovern PE, Zhang J, Tang J, Zhang Z, Hall GR, Moreau RA, Nunez A, Butrym ED, Richards MP, Wang C, Cheng G, Zhao Z, Wang C (2004) Fermented beverages of pre- and proto-historic China. Proc Natl Acad Sci USA 101:17593–17598.
32. McGovern PE (2000) The Funerary Banquet of "King Midas". Expedition 42:21–28.
33. Pliny the Elder (1855) The Natural History. Bostock J, Riley HT (trans) Taylor and Francis, London, Book XIV, Chapter 17.
34. Pliny the Elder (1855) The Natural History. Bostock J, Riley HT (trans) Taylor and Francis, London, Book XIV, Chapter 6.

References

35. Columella LJM (1745) L. Junius Moderatus Columella of Husbandry in Twelve Books and his Book concerning Trees. Millar A (trans) London, UK, Book XII, p 517.
36. Munitis MT, Cabrera E, Rodriguez-Navarro A (1976) An Obligate Osmophilic Yeast from Honey. Appl Environ Microbiol 32:320–323.
37. Reynolds J (1986) The Elder Pliny and his Times. In: French R, Greenway F (eds) Science in the Early Roman Empire: Pliny the Elder, his Sources and Influence. Rowman & Littlefield Publishers, Lanham, Maryland.
38. DiRenzo A (2013) Columella. http://faculty.ithaca.edu/direnzo/gallery/128/?image_id=816. Accessed 27 June 2013.
39. Steinkraus KH, Morse RA (1966) Factors influencing the fermentation of honey in mead production. J Apic Res 5:17–26.
40. Steinkraus KH, Morse RA (1973) Chemical Analysis of Honey Wines. J Apic Res 12:191–195.
41. Partington, JR (1935) Origins and Development of Applied Chemistry. Longmans, Green and Co., London, pp 301–302.
42. Hodges H (1992) Technology in the Ancient World. Barnes & Noble Books, New York, pp 114–117.
43. Broshi M (2007) Date Beer and Date Wine in Antiquity. Palest Explor Q 139:55–59.
44. Murray MA, Boulton N, Heron C (2000) Fruits, vegetables, pulses and condiments. In Nicholson PT, Shaw I (eds) Cabridge University Press, Cambridge, pp 609–655.
45. Samarawira I (1983) Date palm Phoenix dactylifera. Potential source for refined sugar. Econ Bot 37:181–186.
46. Wilkinson JG (1854) A Popular Account of the Ancient Egyptians. Harper & Brothers, New York, Vol. 1, p 55.
47. Pliny the Elder (1855) The Natural History. Bostock J, Riley HT (trans) Taylor and Francis, London, Book XIV, Chapter 19.
48. Yaggy LW, Haines TL (1881) Museum of Antiquity Illustrated. Western Publishing House, p 187.

Chapter 3
Beer

He who knows not beer knows not what is good.—Sumerian proverb [1, 2].

Grain cultivation is thought to date nearly to the initial development of permanent settlements in the Fertile Crescent and there is a strong argument for linking grain cultivation with the civilization of mankind [3]. Cereal grains, which began as relatively unimportant foodstuffs, took on greater significance following a couple simple discoveries. The first was that when grain gets wet, it starts to germinate and becomes sweet tasting [4]. This sweetness is a result of diastase enzymes such as β-amylase, which are present in the germinated grain. These enzymes then convert the starch[1] contained within the grain into maltose sugar (Fig. 3.1) [1, 3–7]. The resulting converted grain is referred to as *malted grain* or simply *malt*. The malting of grains was common practice in the ancient Near East, not only for increased sweetness and palatability, but also for preservation and increased nutritional value [2, 3].

The malting of grains then led to a second important discovery. Grains could be malted and used as a component of soup, or could be heated on their own in water to make a porridge or gruel [4, 7]. When malted porridge was allowed to sit too long, it fermented, and thus provided an entirely new sensation of taste and physical condition [2]. Such fermentation was due to wild yeast present in the air and exposure of the porridge would result in the implantation and growth of yeast cells [7], which would then ferment the sugar from the malted grains into alcohol [4]. The yeast first utilizes the enzyme α-glucosidase to hydrolyze the maltose to two equivalents of glucose (Fig. 3.2). Fermentation of the glucose then occurs in an identical fashion to that previously discussed [1].

---

[1] Starch is the primary form of fuel storage in most plants. The word starch has been said to come from the Old English *stearc* meaning stiff or strong [3].

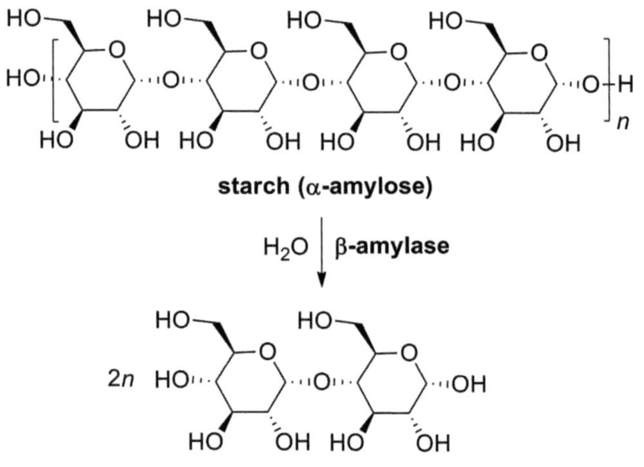

Fig. 3.1 Enzymatic conversion of α-amylose, the simplest form of starch, to the sugar maltose

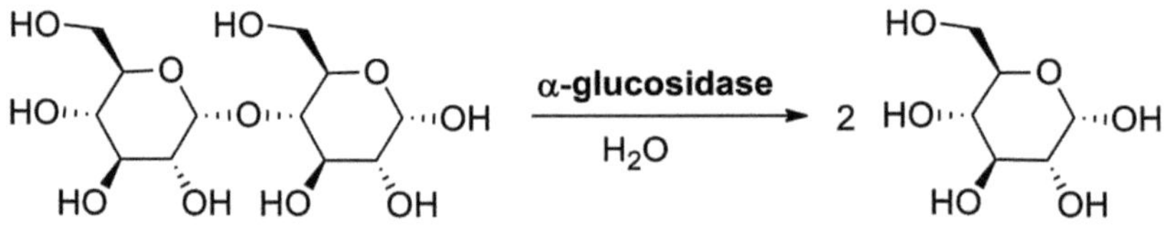

Fig. 3.2 Enzymatic hydrolysis of maltose to glucose

3.1 Beer versus Bread

> Beer is liquid bread.—Justus von Liebig [8].

It is thought that both bread and beer were derived from fermented gruel as discussed above [3, 4]. A thick gruel or dough could be baked to make bread, with the rising of dough resulting from the CO_2 released by fermentation. During baking, the alcohol is cooked off and the effect of the CO_2 evolution results in a much lighter, more digestible, and tastier form of bread [3]. Of course, the fermentation of a thin gruel could then lead to a new type of fermented beverage, beer. Unlike the previous wines, beer could be made from cereal crops, which were abundant and could be easily stored. This allowed beer to be made reliably and in significant quantity [4, 9].

Bread-making and the domestication of grains have long been thought to be interconnected developments, with bread being the primary motivation for the beginning of agriculture. Others, however, have suggested that cereals may have

3.1 Beer versus Bread

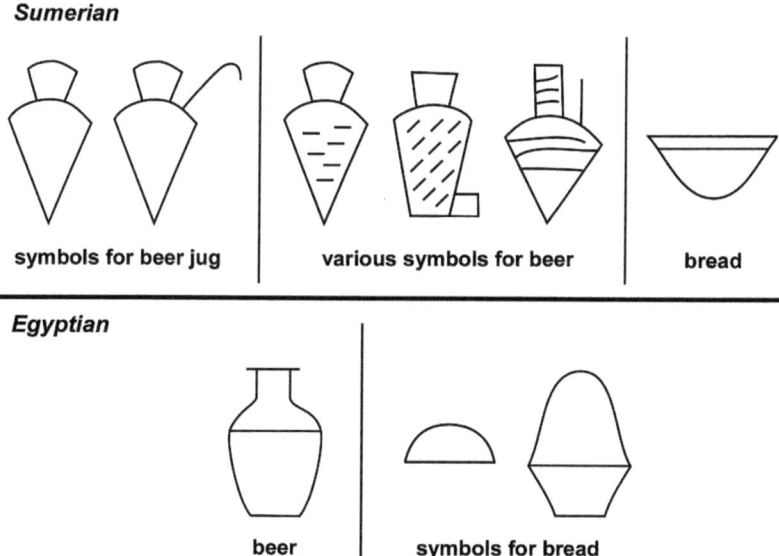

Fig. 3.3 Symbols relating to beer and bread in Proto-Sumerian and Old Kingdom Egyptian scripts [11, 14]

been domesticated initially for the making of beer rather than bread [7, 10–13]. Originally proposed by J. D. Sauer (University of Wisconsin), this question was formulated by Robert J. Braidwood (University of Chicago) in 1953 as follows [10]:

> Could the discovery that a mash of fermented grain yielded a palatable and nutritious beverage have acted as a greater stimulant toward the experimental selection and breeding of the cereals than the discovery of flour and bread making? One would assume that the utilization of wild cereals (along with edible roots and berries) as a source of collected food would have been in existence for millennia before their domestication (in a meaningful sense) took place. Was the subsequent impetus to this domestication bread or beer?

After significant debate, a narrow margin still favored bread as the primary motivation for agriculture, but arguments for the contributions of beer still remain [7, 10, 11]. Even so, it has been stated that it is likely that the preparation of unleavened bread and beer (and possibly even the accidental leavened bread) occurred before agriculture. While no direct evidence of such has been found, the large number of wild grains suitable for use in beers, the simple methods involved, and the number of groups who prepare beers all support such a possibility [10].

As will be discussed further below, the connection between bread and beer is further strengthened by the fact that some early beers used bread as an ingredient, suggesting that bread was a means of storing the raw materials for beer [1, 12]. Early Mesopotamian and Egyptian notational systems include signs for both 'bread' and 'beer' [11, 14] (Fig. 3.3) and references to the use of bread in brewing by both cultures have led to much debate among archaeologists and historians as to the proper chronology of bread and beer. Some have suggested that bread must be a derivative of beer making, while others argue that bread came first and was later used as a component in beer [4]. An argument in support of beer being the initial cereal product was given by Paul C. Mangelsdorf (Harvard University) as follows [10]:

> A fairly good case might be made for an earlier utilization of cereals for brewing than for bread-making, not because thirst was a stronger motive than hunger, but because the earliest grains available were more suitable for beermaking than for bread. The earliest cereals of the Near East—the wild wheats, *Triticum aegilopoides, T. thaoudar*, and *T. dicoccoides* and their cultivated counterparts, *T. monococcum* and *T. dicoccum*, and species of barley—are all characterized by the adherence of the glumes (husks or chaff) to the grains after the grains have been removed from the heads. Such glume-enclosed grains are, without additional processing, virtually useless for bread-making but all can be used for brewing and one, barley, is the cereal *par excellence* for this purpose.

A group of researchers from France attempted to address this debate in 2007, as well as others, by investigating the genetic diversity of *Saccharomyces cerevisiae* via a large-scale evaluation of various yeast populations, including strains associated with bread, beer, wine, and sake [15]. The genetic differences between yeast groups observed in the study suggest an ancient divergence leading to local natural populations from which multiplication may have been favored by humans. Comparing genetic relationships of the bread, beer, and wine strains studied, the authors proposed that the main group of bread strains resulted from a tetraploidization event between an ale beer strain and a wine strain. Such a conclusion would thus imply that bread technology appeared after the production of both beer and wine [15].

3.2 Barley Beer

While fermented beverages can be produced from a number of grains, beer traditionally comes primarily from barley (*Hordeum vulgare L.*, Fig. 3.4). Barley was one of the first of the grains to be domesticated, if not the very first, with this domestication commonly thought to have first appeared in or near Mesopotamia as early as 8000 BCE [1, 12]. It has been suggested, however, that botanical evidence points to the first domesticated barley being not in the Near East but the Far East, possibly Tibet [10]. Domestication of barley came later in Egypt, with its cultivation dating back to at least the sixth millennium BCE [16]. The brewing of beer from barley is thought be linked to nearly the beginnings of agriculture [10–12] and is believed to date back to approximately 6000 BCE [1, 3]. For this reason, barley and the date palm are thought to provide us with the first direct evidence that both the Mesopotamians and the Egyptians were making fermented drinks [5].

Of the common grains, barley produces the greatest amount of diastase enzymes upon germination, with wheat being second (about 50 % that of barley) [1, 4, 7, 17]. As a consequence, barley also produces the most maltose during malting and provides a near ideal grain for the production of beer. Once the crucial

3.2 Barley Beer

Fig. 3.4 *Hordeum vulgare L.* (Botanical print (dated 1885); barley ear photo by Dag Endresen 2004; barley field photo by Geert Orye 2005)

discovery of barley beer had been made, its quality was improved through trial and error. The more malted grain there is in the original gruel and the longer it is left to ferment, the stronger the beer. More malt means more sugar and a longer fermentation means more sugar is converted to alcohol [4]. As cooking bread results in desirable chemical changes, it was ultimately found that cooking the malted grain also contributes to the strength of the beer. The malting process converts only about 15 % of the starch in barley to sugar [4]. However, the starch-converting enzymes become active at higher temperatures (60–70 °C) and turn even more of the starch to the necessary sugars, thus allowing the production of even greater amounts of alcohol [1, 3, 4]. The result of this process is referred to as the *wort* and fermentation of the cooled wort results in the final beer [1].

The modern use of the word beer refers to a barley beverage that is brewed in the presence of hops. The addition of hops gives beer a more bitter taste, but also improves its storage properties due to the preservative properties of hops [18]. The use of hops, however, was not introduced until the 15th century, usually attributed to the Dutch [1, 4]. Others point to either the Finnish, Letts, or Slavs as the first to use hops [18, 19]. As a result of this late introduction of hops, the term beer in a historical context refers to all fermented products of barley and is not limited to those beverages that contain hops.

Beer maintained a widespread popularity in the ancient Near East. For millennia it was the favorite drink in Egypt and Mesopotamia, fit not only for mortal consumption but deemed worthy of the gods as well [2]. The Babylonians knew of at least seventy varieties of beer, and texts of the Hittites of Anatolia suggest that beer was second in importance only to bread [1, 2]. Beer was also popular among the Phrygians and Armenians [2, 20], as illustrated by the description of Armenian beer by Xenophon[2] in his *Anabasis* [21]:

> ...and wine made from barley in great big bowls; the grains of barley malt lay floating in the beverage up to the lip of the vessel, and reeds lay in them, some longer, some shorter, without joints; when you were thirsty you must take one of these into your mouth, and suck. The beverage without admixture of water was very strong, and of a delicious flavour to certain palates, but the taste must be acquired.

In fact, it has been said that beer only suffered a decline in popularity after the Greeks Hellenized the Near East, possibly because the Greeks championed wine as a replacement [2].

3.3 Beer versus Wine

Of course, one cannot properly discuss the history of beer without addressing its often debated chronology in relation with the other most common representative of fermented beverages, i.e. grape wine. A complication in the longstanding debate as to which came first, barley beer or grape wine, are the specific terms themselves, *beer*[3] and *wine*. In their modern use, the term beer is always associated exclusively with beverages from fermented grains, while wine is almost always associated with those beverages obtained from fermented fruits. Historically, however, these modern associations were much less distinct and older terms that we tend to correlate to the more modern beer or wine could often be used interchangeably. Examples of this include references in the Old Testament that could denote either beer or grape wine and the fact that date wine was often referred to using terms that correlate better to beer than wine [9]. Even the Old English word for beer,

---

[2] Xenophon was born 431 BCE [20, 21], near Athens, and was a pupil of Socrates [21]. As a soldier, he fought under the Spartan king, and was exiled from Athens. Sparta gave him land and property in Scillus, where he lived for many years before having to move once more, to settle in Corinth [21]. His work *Anabasis* describes the travels of Greek mercenaries to Persia and their return to Thrace following the battle of Kunaxa (401–399 BCE) [20, 21]. He died in 354 BCE [21].

[3] The modern word *beer* is believed to derive from the Old English *beor,* a word whose own origin is ambiguous and much-disputed [22]. Some maintain that it derives from the Latin *bibere* "to drink". Others suggest a relationship with the Old Low German word *beo*, which derives from the Proto-Germanic **beuwo-* "barley." The Greeks, however, referred to beer by the word *zythos* (sometimes given as *zythus* or *xythus* [19]), while the Egyptians called beer either *haqu* or *heqa* [8].

beor, was used to refer to both beer and mead (honey wine) [22]. This issue can also been seen among various classic Greek texts as illustrated by the quote given by Xenophon above and the following by Herodotus, both of which refer to beer as a barley-wine [8, 23]:

...they use habitually a wine made out of barley, for vines they have not in their land.

Greek writers would also use the term barley-mead, in addition to barley-wine, to refer to beer [8].

Even ignoring the complicating factor of imprecise language, others have previously noted that it is quite difficult to ascertain with any certainty whether wine or beer came first [17]. Both barley beer and grape wine are thought to date back to approximately 6000 BCE [1, 3], although it has been proposed that a "wine culture" was established by at least 7000 BCE [24]. To support claims that wine predated beer, some have pointed to the common mention of wine in the Bible, but the lack of such mentions of beer [17]. However, it should be noted that the usual word used in the Old Testament is *šêkar*, which refers simply to an intoxicating drink. While it is thought that this usually means wine, it could also perhaps mean beer [25]. Others have instead quoted the opinions of various classic Greek authors which also favor wine over beer [17]. However, the strongest argument for wine over beer has been the fact that the principal yeast species responsible for fermentation, *Saccharomyces cerevisiae*, occur naturally in damaged grapes [26, 27].

The combination of the fact that *Saccharomyces cerevisiae* occurs naturally in the grape, and doubts about the airborne transmission of either the organism or its spores, has led to the argument that a practical understanding of the use of *Saccharomyces cerevisiae* in fermenting grapes must have preceded its use for grain-based foods and beverages such as bread or beer. [26, 27]. An extension of this argument is that the yeasts from the grapes were then initially used to ferment the later cereal products, with the yeasts eventually evolving into specific strains over centuries of human selection [26]. Further support for this argument has been reference to a line in the 1800 BCE hymn to the Sumerian goddess of brewing, the *Hymn to Ninkasi* [12, 28], that implies the addition of wine, grapes, or raisins to the sweetwort prior to fermentation [12, 26].

As previously described, the genetic diversity of *Saccharomyces cerevisiae* via a large-scale evaluation of various yeast populations was performed by a group of researchers from France in 2007 [15]. One of the primary questions addressed by the study was whether evidence could be found that beer yeast strains evolved from wine strains as proposed by the above argument. Comparison of the genetic relationships of a large number of various strains, however, actually revealed that the beer strains were quite poorly related to wine yeast, leading the researchers to state [15]:

However, as beer strains are obviously far from wine yeast, our results do not support the classical hypothesis of wine technology as an origin for beer.

Without more substantial evidence one way or the other, it is really just not possible to state explicitly whether either barley beer or grape wine preceded the

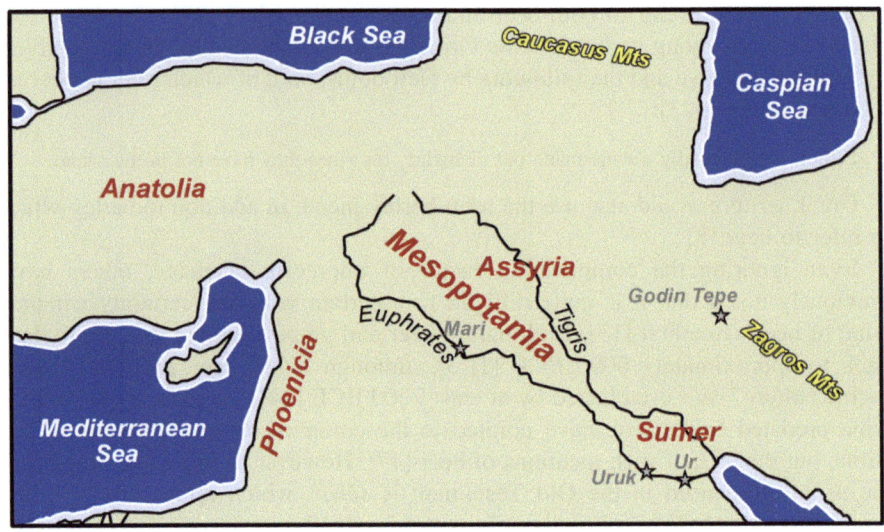

Fig. 3.5 Mesopotamia, ca. 4300–2000 BCE

other. What is known, however, is that beer was the predominate drink during the early periods of Mesopotamia and Egypt [2] and the popularity of wine only increased with the later influx of Greek populations [17]. As such, it is quite reasonable to propose that both barley beer and grape wine were introduced during the same initial time period, but that due to the easy accessibility of large amounts of grain, beer became more widespread and common at an earlier time period than grape wine. Others have come to similar conclusions, citing that the restriction of wine production to a limited number of areas in both Egypt and Mesopotamia during the Bronze Age undoubtedly had much to do with the agricultural conditions necessary for large-scale, sustained production. In comparison, as the basis for beer, barley was the ubiquitous cereal staple throughout the archaeological and cultural records of these two regions [17, 29].

3.4 Beer in Mesopotamia

The consumption of beer is believed to have first appeared in or near Mesopotamia as early as ∼6000 BCE [1, 3]. The name Mesopotamia derives from the Greek meaning "[the land] between the rivers", which referred to its placement in the floodplain between the Euphrates and Tigris rivers (Fig. 3.5) [9]. Silt spread by flooding from the rivers resulted in the creation of fertile soil, and once early Mesopotamians learned how to irrigate this land by digging canals to carry water from the rivers, this area became nearly perfect for the development of agriculture. Such efforts included the domestication of cereals such as barley, which as

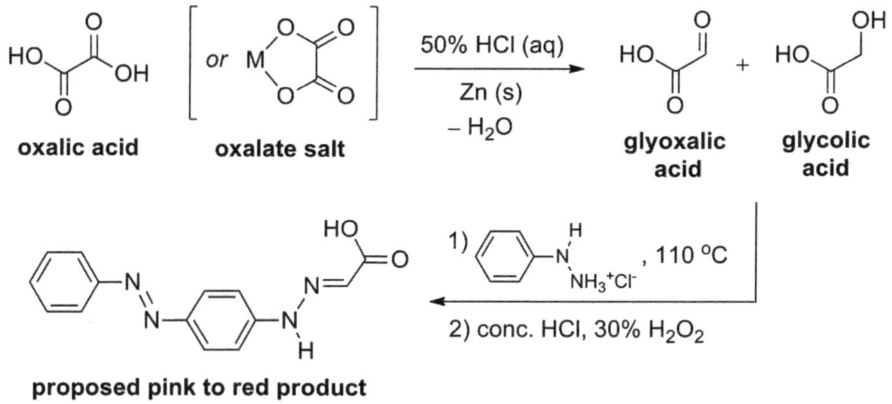

3.4 Beer in Mesopotamia

Fig. 3.6 Feigl spot test for oxalate

discussed above, is the preferred grain for the production of beer [30]. With such a rich location, it is not surprising that southern Mesopotamia became the home to the early Sumerian city-states, collectively thought to be the oldest civilization in the world, which emerged ∼3400 BCE [11, 13]. Of course, the preferred fermented beverage of the ancient Sumerians was beer [7, 30]. The popularity of beer was not just limited to the Sumerians, though, with beer consumption occurring throughout Mesopotamia during all eras and by all social classes, including women [11].

As is often the case, the oldest and most internally diverse societies typically provide some of the most fragmentary evidence for documenting their detailed history and practices. As such, the history of alcoholic beverages in Mesopotamia can be more complex than that of later regions such as Egypt or the Levant. However, researchers have been able to collect various pieces of evidence to help create a strong circumstantial case for the place of alcoholic beverages in early Mesopotamian history [6].

Unlike wine, there have been limited chemical archeological findings reported in relation to beer. The only real chemical evidence relating to beer in Mesopotamia comes from the analysis of a pale yellowish residue found in grooves of a jar from Godin Tepe (Fig. 3.5), a site of strong Lower Mesopotamian influences in the Zagros mountains in Iran [9, 13, 30, 31]. The double-handled jug was recovered from one of the same rooms containing a wine jar, and dates to the Late Uruk period (∼3500–2900 BCE) [31]. Based on contextual evidence, the authors strongly felt that this vessel served as a beer container.

The yellow residue was subjected to a Feigl spot test for oxalate [32] (Fig. 3.6) which gave positive response for oxalate, presumably calcium oxalate [13, 30, 31]. Calcium oxalate is a principal component of "beerstone," an insoluble deposit[4]

---

[4] Its solubility in water is 6 mg/L at 18 °C [26, 27]; K_{sp} ($Ca(C_2O_4) \cdot H_2O$) = 2.34×10^{-9} [33].

which accumulates on the inner surfaces of fermentation and storage vessels and has thus been used to support the evidence of beer [11, 13, 30, 31]. To further support the presence of calcium oxalate, tests were also carried out on scrapings from an Egyptian New Kingdom blue-painted jar (which tomb paintings and reliefs suggested was intended for beer) [11, 30, 31], beerstone from a modern brewer's vat, and pure calcium oxalate. In all cases, similar test results to the yellow residue were obtained [30, 31]. The presence of oxalate in the examined jar was thus attributed to contents of fermented grains [30] and reported to confirm the archaeological and pictographic evidence that the vessel was a beer container [31].

Calcium oxalate, however, is a very simple chemical species and can result from a number of sources. In is known in three primary forms: the monohydrate $Ca(C_2O_4) \cdot H_2O$ (the mineral whewellite), the dihydrate $Ca(C_2O_4) \cdot 2H_2O$, and the polyhydrate $Ca(C_2O_4) \cdot (2 + x)H_2O$ (the mineral weddelite) [34]. Microcrystals of the polyhydrate are found in some plants, as well as in the form of a sediment in cooled urine. In fact, it is a major constituent of human kidney stones [34]. Likewise, microcrystals of the monohydrate are common in higher species of plants and the authors themselves point out that oxalates occur naturally in relatively large amounts ($\sim$5–10 % by fresh weight) in spinach and rhubarb, species of which grow in the Iranian highlands today [30, 31]. In addition, both the monohydrate and polyhydrate are known mineral species [34]. As such, it has been pointed out that deposits of oxalate salts on the pottery analyzed could just as well result from their being buried in oxalate-rich soil [11]. Another issue is that while the spot test utilized is quite sensitive (sensitive to 10^{-6} g), Feigl warns that the presence of either glyoxalic or glycolic acid in the sample will give a false positive [32]. In order to remove this possibility, the sample must first be precipitated as the insoluble calcium oxalate salt and washed well with water in order to remove the presence of any other acids or salts [32]. Assuming that the sample isolated from the pottery sherd was beerstone, washing first with water prior to the spot test could remove the possibility of such a false positive, but no such pretreatment of the sample was reported [30, 31].

In efforts to rule out various environmental oxalate contributions as discussed above, the exterior of the "beer" jar and the interior of one of the wine jars from the same sight were subjected to Feigl spot tests, both of which gave negative results [30, 31]. Even so, it is not possible to reliably say that oxalate detected is from beerstone and it is recognized that there are currently no reliable bio-markers available for confirming the presence of beer in ancient residues [11]. Others have gone even further to state that ancient organic residues in general can be challenging and unreliable samples for analysis [11, 29], a topic which will be discussed in more detail in the following chapter.

While reliable chemical evidence of Mesopotamian beer technology is limited, written records have provided a fair amount of useful information. Tablets have been found dating back to $\sim$2800 BCE containing recipes for beer [7, 11, 13], with some sources citing tablets of even older time periods [35, 36]. The code of Hammurabi, dated to the 18th century BCE, details laws codified by King Hammurabi of Babylon and includes laws specifically aimed at beer parlors. Prescribed by law were stiff penalties for vendors who sold weak beer, overcharged their customers (death by drowning), permitted criminals to frequent their establishments (execution), or allowed political conspiracies to be created on their premises [1, 12]. Archaic texts from the Sumerian city of Uruk (Fig. 3.5) also list very large quantities of goods manufactured, stored, and distributed, including vessels and beer [6, 13]. One such administrative archive records the production of at least eight different types of beer [6].

The recipes referred to in these ancient tablets recovered are generally not true detailed recipes and some are too fragmented to really be of much use. However, many of these are temple accounts which document the type and amount of grain issued to brewers and the amount of beer received in return. Thus, they at least provide an idea of the ingredients of the beer and the ratios involved [11]. Perhaps the best studied of these records is the *"Hymn to Ninkasi"*, which was written about 1800 BCE and has been found on tablets at Nippur, Sippar, and Larsa [1, 11–13]. As introduced above, this hymn is in praise of the Sumerian goddess of brewing and gives the basic outline of the brewing process [1, 7, 12, 13, 28]. The hymn was originally translated by Miguel Civil (Oriental Institute of the University of Chicago) in 1964 [28]. Civil, however, revisited and refined the translation again in 1991 [11, 12]. Civil's refined translation of the hymn is as follows [12, 28]:

> Borne *by* the flowing water [...],
> Tenderly cared for by Ninhursag,
> Ninkasi, borne *by* the flowing water [...]
> Tenderly cared for by Ninhursag.
>
> Having founded your town by the sacred lake,[5]
> She finished its great walls for you,
> Ninkasi, having founded your town by the sacred lake
> She finished its great walls for you.
>
> Your father is Enki, Lord Nudimmud,
> Your mother is Ninti, the queen of the sacred lake.[6]
> Ninkasi, your father is Enki, Lord Nudimmud,
> Your mother is Ninti, the queen of the sacred lake.
>
> You are the one who handles the dough [and] with a big shovel,
> Mixing in a pit, the bappir with sweet aromatics.
> Ninkasi, you are the one who handles the dough [and] with a big shovel,
> Mixing in a pit, the bappir with [date]-honey.[7]
>
> You are the one who bakes the bappir in the big oven,
> Puts in order the piles of *hulled* grain.
> Ninkasi, it is you who bake the bappir in the big oven,
> Puts in order the piles of *hulled* grain.

---

[5] Originally translated in 1964 as *"founded your town on 'wax',"* [28].
[6] Originally translated in 1964 as *"the queen of the abzu"* [28].
[7] Originally translated in 1964 as *"with sweet aromatics"* [28].

You are the one who waters the malt set on the ground,
The *noble* dogs keep away even the potentates.[8]
Ninkasi, you are the one who waters the malt set on the ground,
The *noble* dogs keep away even the potentates..

You are the one who soaks the malt in a jar,
The waves rise, the waves fall.
Ninkasi, you are the one who soaks the malt in a jar,
The waves rise, the waves fall.

You are the one who spreads the cooked mash on large reed mats,
Coolness overcomes,
Ninkasi, you are the one who spreads the cooked mash on large reed mats,
Coolness overcomes,

You are the one who holds *with both hands* the great sweetwort,
Brewing [it] with honey [and] wine.
[You the sweetwort to the vessel].
Ninkasi, [...],
[You the sweetwort to the vessel].

The fermenting vat, which makes a pleasant sound,
You place appropriately on [top of] a large collector vat.
Ninkasi, the fermenting vat, which makes a pleasant sound,
You place appropriately on [top of] a large collector vat.

When you pour out the filtered beer of the collector vat,[9]
It is [like] the onrush of the Tigris and the Euphrates.
Ninkasi, you are the one who pours out the filtered beer of the collector vat,
It is [like] the onrush of the Tigris and the Euphrates.

The hymn has been interpreted as a linear description of brewing and seems consistent with the general steps discussed above for barley beer. However, it has been pointed out that given that some passages of the text are obscure, the translation is influenced to a considerable extent by knowledge about modern brewing technology. As such, opinions on the meaning of various points differ considerably in the scholarly literature [13]. One point of significant interest is the fact that the hymn does not make any reference to the process of malting [11], although one stanza has been translated to include the word malt [12, 28]. This has led some to view that the preparation of malt was not a part of Mesopotamian brewing technology [11]. Others, though, have stated that flour from malted barley was probably added later in the process [12]. Interpretation of the hymn as a beer recipe was finally put to the ultimate test by Solomon Katz (University of Pennsylvania) and Fritz Maytag (Anchor Brewing Company) who used the hymn as a basis for the modern brewing of a Sumerian beer [11, 12]. The resulting beer, appropriately called Ninkasi, yielded an alcohol content of 3.5 % and was

---

[8] Originally translated in 1964 as *"waters the earth-covered malt, The noble dogs guard (it even) from the potentates."* [28].

[9] Originally translated in 1964 as *"You are the one who pours out the filtered beer"* [28].

described to have a dry flavor lacking in bitterness, tasting similar to a hard apple cider with the fragrance of dates [12].

From the available evidence, it is generally believed that beer in Mesopotamia was prepared via the following steps and processes. The first stage was preparing the malt. Here the barley was soaked in water, thus allowing it to germinate for a short time [5, 11] and then dried, either by laying it out in the sun for about three weeks, or by lightly heating it in a kiln. This produced "green" malt, which was then further kilned at a slightly higher temperature to produce "cured" malt [11].

It is then thought that the malt was crushed by pounding, sieved to remove husks, etc., and then either stored or converted into dough and baked in a domed oven [5, 11]. These small malt loaves, which were called *bappir* in Sumerian and *bappiru* in Akkadian, normally had various aromatic components incorporated into them and represented one of the main types of raw material for fermentation [11]. Some records indicate that bappir was kept in government storehouses and was only eaten during food shortages. Thus it was not considered a foodstuff, but a convenient way to store the raw material for making beer [4, 12].

These malted loaves were then crumbled and added to water (possibly with more grain), heated, and mixed to make the mash [5, 11]. The mash was then allowed to cool, and sweeteners such as honey or date juice were often added. Once fermentation started, the mash was regularly stirred before being transferred to another vessel upon the completion of fermentation. This vessel was a clarifying vat that permitted some degree of clarification by sedimentation [11]. The final beer was then drawn off and put into stoppered storage or transport jars to prevent the further fermentation which would make the drink acidic [5, 11].

Mesopotamian beer from the earlier periods was thought to be unfiltered and evidence from ancient Phrygia (8th–7th century BCE) in Anatolia indicates that brewing left barley husks and other solid matter in the beer. To avoid this material, such unfiltered beer was drunk either through a straw from an open vessel or from a jug with a spout attached to the main body through a sieve [1, 7]. This is illustrated by iconographic evidence (3rd millennium BCE) that depict banquet scenes in which groups of seated individuals are drinking from vessels through large tubes or straws (Fig. 3.7) [4, 37, 38]. These scenes appear in both glyptic and decorative arts, and drinking vessels and straws have been found in elite burials, most notably the Royal Cemetery at Ur [11]. Historical evidence also indicates that filtered beers were brewed and it is believed by some that the move to filtered beers constitute a major change in brewing technology during the mid-3rd millennium BCE [11].

Based on ceramic evidence from Mesopotamia, it is thought that production and consumption of alcoholic beverages increased throughout the 4th and 3rd millennia [6]. Comparison of 3rd and 2nd millennium BCE texts from Mari show that while beer rations are commonly referenced in the 3rd millennium, such references occurred rarely in the 2nd millennium and were generally replaced by documentation of the rigorously controlled royal distribution and consumption of wine [6].

Fig. 3.7 Sketch of a Mesopotamian seal depicting drinking from tubes, ca. 2500 BCE (Image of the original stone seal is available as image #158 in Ref. [38])

3.5 Beer in Egypt

Bread and beer were dietary staples throughout ancient Egypt [39–44] and evidence for the production and use of beer in Egypt extending back to the Predynastic era (5500–3100 BCE) has long been known [17, 19]. In fact, Greek writers have credited the Egyptians with having invented beer, although modern historians believe that there is no credible foundation for this assertion [17]. Written records of the Early Dynastic period (3100–2686 BCE) also indicate that beer was very important and must have been a well-established feature of the culture and period [17].

Many types of beer were manufactured in ancient Egypt, with the majority of these based on the fermentation of malted barley and/or emmer wheat (*Triticum dicoccum* Schübl.) [6, 8, 16, 39–42]. Emmer wheat is now fairly rare, but was one of the original domesticated crops which was dominant throughout much of the Old World [42]. On the basis of documentary evidence, it is believed that some varieties were also flavored by the addition of dates, lupins, fruits, or wines [6, 8, 19, 39–41]. Such beers were drank by all social classes in Egypt, from the Pharaoh downwards [17, 40, 42, 43], and beer was inextricably woven into the fabric of daily existence [17]. In Alexandria, however, beer was said to be used primarily by the lower classes [19]. Egyptian beer was thought to be drunk from flat dishes without feet, and at least some varieties were said to be strongly intoxicating. The intoxicating quality of the beer was thought to be due to a demon, and a papyrus of 1200 BCE contains a magic formula to get rid of its effects [19].

In contrast, it has been thought by some that Egyptian beer would have been drunk daily as a highly refreshing and more reliably potable substitute for water, which was not always noted for being especially clean or hygienic [16, 35, 41]. As such, it is thought that such beers brewed for everyday drinking would not

3.5 Beer in Egypt

necessarily have been very alcoholic and would thus have had a relatively short shelf-life. If so, this would have necessitated frequent brewing and immediate consumption with no long-term storage [17].

Evidence for Egyptian beer production and methods includes excavations from two Pre-dynastic sites at Abydos and Hierakonpolis. These excavations uncovered what is considered to be the remains of breweries, including double rows of large vats, supported by distinctive firebricks [6, 40, 44]. A hardened black residue was recovered from the vats in which intact or fragmentary cereal grains were embedded and initial chemical analysis suggested that the product could have been beer [40]. Later two additional samples from these vats, characterized as amorphous charred residues, were analyzed in efforts to determine their composition. Macro- and micro-scopical examination were reported to reveal the presence of intact remains of wheat (*Triticum dicoccum*) and barley (*Hordeum vulgare*), as well as fragments of dates (*Phoenix dactylifera*) and pips of grapes (*Vilis vinifera*).

The residues were then subjected to chemical analysis, and while the majority of the residue was determined to be ash, three water-soluble fractions were separated via ion exchange chromatography, which were reported to be sugars, amino acids, and organic acids [44]. The isolated sugar fraction was then O-silylated and analyzed by gas chromatography (GC), the carboxylic acid fraction treated with diazomethane and analyzed by gas liquid chromatography (GLC), and the amino acid fraction analyzed by high pressure ion exchange chromatography [44]. The sugar fraction was reported to contain xylose, mannose, galactose, glucose, galacturonic acid, and glucuronic acid (Fig. 3.8), as well as three unknown chemical components, while the acid fraction was reported to contain oxalic, succinic, tartaric, lactic, and malic acids (Fig. 3.9).

Lastly, the amino acid fraction was reported to consist of $\sim 30\,\%$ ammonia, with the remainder attributed to 16 different amino acids. The highest percentages reported were for aspartic acid, threonine, glutamic acid, glycine, and lysine [44]. It should be pointed out that at no point in this study did the authors use any chemical method to confirm the chemical identity of any of the species reported and all conclusions were based only on retention times in relation to commercial standards. Nevertheless, this did not keep the authors from claiming that these results confirmed that vat residue samples were fermentation products of cereal grains and fruit material [44].

Beer sediment had also been reportedly found from jars at Abadiyeh, a Pre-dynastic cemetery on the east bank of the Nile in upper Egypt, as well as at Naqada, one of the largest Pre-dynastic sites in Egypt, situated on the west bank of the Nile [17, 44]. In addition to the concerns about the rigor of the analysis methods reported above, Delwen Samuel (Kings College London) has stated the following concerning these specific residue analyses [40]:

> The analytical chemical work published to date is unconvincing because the effects of degradation of ancient organic molecules (and therefore the possible presence of molecules that were not part of the original contents) has not been considered. Also, the possibility that micro-organisms contaminated the organic remains after abandonment has not been eliminated.

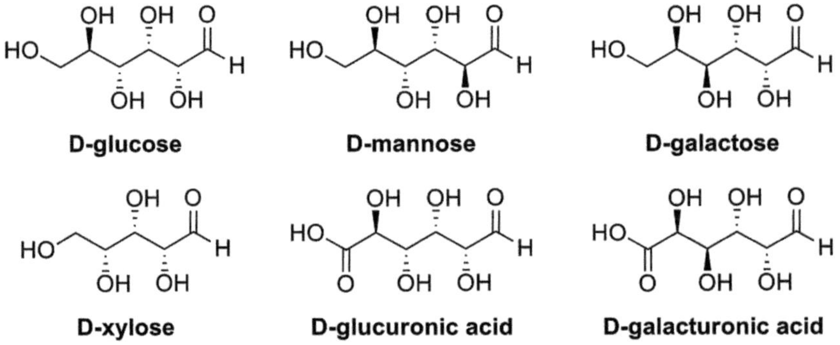

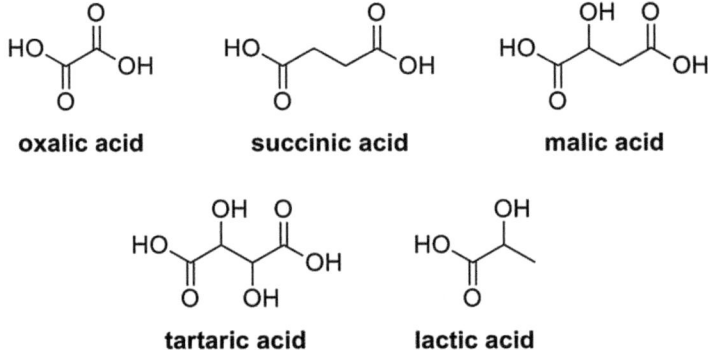

Fig. 3.8 Sugars reported to be detected in residue from Hierakonpolis

Fig. 3.9 Organic acids reported to be detected in residue from Hierakonpolis

One of the earliest written accounts of the Egyptian methods for brewing beer was given by the alchemist Zosimus of Panopolis[10] in ca. 300 CE. A 1922 translation of this account states [17]:

> Take fine clean barley and moisten it for one day and draw it off or also lay it up in a windless place until morning[11] and again wet it six hours. Cast it into a smaller perforated vessel and wet it and dry it until it shall become shredded and when this is so pat it (i.e. shake, or rub) in the sun-light until it falls apart. For the must is bitter. Next grind it and make it into loaves adding leaven, just like bread and cook it rather raw and whenever the

---

[10] Zosimus of Panopolis was an alchemist, Gnostic, and Hermetic philosopher, who lived around 300 CE (3rd–4th century CE) [45–49]. He was one of the early authors on alchemy that was not thought to be a false attribution or pseudonym, and he wrote the earliest known alchemical encyclopedia, *Cheirokmeta*, a work of 28 books [45–48]. Although he is always associated with Panopolis (ancient Akhmim), which is in Upper Egypt [17, 40, 47], he wrote and taught in Alexandria [17, 45].

[11] An earlier 1814 translation interpreted this line quite differently, giving "*...spread it for a day in a spot where it is well exposed to a current of air.*" [8].

3.5 Beer in Egypt

loaves rise, dissolve sweetened water and strain through a strainer or light sieve. Others in baking the loaves cast them into a vat with water and they boil it a little in order that it may not froth nor become luke-warm and they draw up and strain it and having prepared it, heat it and examine it.

Early Egyptian brewing activities have also been interpreted from the numerous artistic records of ancient Egypt (various wall-reliefs and paintings, models and statues) [17, 39–42]. Although there is vague general agreement, there has been no definite consensus of opinion on the precise brewing methods applied in ancient Egypt and most interpretations vary in aspects of production [41, 42]. However, the following general process has been proposed for the Egyptian brewing methods based on these artistic sources. First a calculated quantity of grain (primarily barley [39]) would be moistened, placed into a mortar, and ground. Yeast would then be added and the mixture worked into dough. This dough would then be placed into earthenware vessels heated over a slow-burning fire which would partially bake the enclosed dough [17, 19, 40, 41]. The partially baked loaves were removed from their vessels, crumbled, and allowed to soak in a large vat of water, where fermentation would occur [17, 19, 36, 39–41]. It has also been reported that honey was probably added in some cases [8]. This addition would produce a beer higher in alcohol, which would be consistent with the highly intoxicating nature used by some ancient authors to describe Egyptian beer [19]. This would also then make the term barley-mead used by some authors a fairly apt description [8].

When fermentation was deemed to be complete, the porridge-like mixture would be transferred to a woven basket-work sieve, where it would be kneaded by hand and the liquid fraction forced through into a large, wide-mouth jar situated below. From this initial large jar, the filtered beer was poured into the final beer jars, which are normally shown as being held in some sort of rack. Once filled, the beer jars were sealed with a ball of mud in the neck of the jar [17, 19]. It has been reported, however, that as the artistic evidence is not completely clear, notable discrepancies in these interpretations persist [39, 41, 42]. For example, the use of malt has been much debated [19, 41]. Another issue that has been noted is confusion over the extent to which brewing practices changed over time [40, 42].

As can be seen in both of the descriptions above, these methods are usually described as utilizing partially baked barley loaves in a manner similar to the Mesopotamian methods. As such, it is not surprising that it is thought that Egyptian bakers often doubled as brewers [2]. As discussed for beer above, traditional descriptions of ancient Egyptian bread making also relies on artistic depictions and written sources. However, it has been noted that available written sources contain little data on bread ingredients [39].

In order to contribute a new perspective in the investigation of Egyptian brewing and baking methods, Delwen Samuel reported the systematic analysis of desiccated bread loaves and cereal residues by optical and scanning electron microscopy (SEM) in 1996 [39, 41]. The bread analyzed derived from sites dating to 2000–1200 BCE, while the residues were from two sites: Deir el-Medina (1550–1307 BCE), on the west bank of the Nile [17, 39–41], and the Workmen's village of Amarna ($\sim$1350 BCE) in middle Egypt [39–41]. Nearly 70 loaves and more than 200 residue samples of pottery vessels were first examined optically to provide data about ingredients and their quantities, unintended inclusions, and texture [39]. The texture of both the bread and residues ranged from fine to coarse. Unlike bread, however, large amounts of chaff were found to be a common feature in the cereal residues, with coarsely shredded husks occurring in substantial quantities in large jars. In thin coatings, husk slivers are often embedded along with small bran pieces.

The ancient Egyptian brewing sequence was then investigated by SEM examination of the beer residue microstructure [39–42]. This examination revealed a wide range in starch morphology, from undistorted but pitted granules to thoroughly fused starch, which was attributed to the use of a two-part brewing process by the Egyptians—one consisting of coarsely ground, well-heated malt or grain and another consisting of unheated malt [39–42]. Samuel points out that it seems very likely that the ancient Egyptians used a variety of techniques to kiln their germinated grain or to process unsprouted grain destined for brewing, which might account for many of the various named types of ancient Egyptian beer [41]. Samuel goes on to stress that the microstructural data revealed by the SEM analysis did not match the accepted use of lightly baked bread for brewing as described in the generally accepted process above. Such partially cooked malt loaves would contain unblistered, undistorted, and pitted starch granules, but would not contribute the extensive quantities of well-fused starch found in the residues. On the other hand, if quite moist loaves had been baked long enough to create large amounts of completely merged starch, few unblistered and undistorted starch granules should have remained [39, 40].

An attempt to gain insight into the fermentation stage of brewing was then accomplished on the basis of the presence or absence of yeast in the different residues investigated. Nearly all the large cereal-based contents of whole jars revealed no traces of yeast, but large pieces of chaff and coarse fragments of grain. The high proportion of chaff and lack of yeast were attributed to the contents consisting of spent grain (i.e. residues after rinsing sugars, dextrins, and starch from processed malt). From this the author proposed that fermentation was initiated in the rinsed sugar- and starch-rich liquid obtained after straining out the bulk of the cereal husk [39, 40].

As can be seen, the conclusions of Samuel [39–41] are in stark contrast with those determined from both artistic depictions and later period written sources [17, 36, 39]. However, to date, no single study has been able to provide a definitive process for the brewing methods of ancient Egyptians. As such, the existence of multiple proposed methods indicates that our current understanding of these ancient processes is far from complete and there is still ample room for additional research [41]. Of course, it is also quite possible that the Egyptians did not use a single method for the production of beer. This thought was proposed by John Arnold in 1911, reasoning that it was well known that the Egyptians produced a number of different beers and while the differences between these beers could be the result of different ingredients, it is also quite reasonable to speculate that they could be the result of multiple brewing processes [8].

References

1. Lambert JB (1997) Traces of the Past. Unraveling the Secrets of Archaeology through Chemistry. Addison-Wesley, Reading, MA, pp 134–136.
2. Sams GK (1977) Beer in the City of Midas. Archaeology 30:108–115.
3. Hornsey IS (2003) A History of Beer and Brewing. The Royal Society of Chemistry, Cambridge, pp 9–20.
4. Standage, T (2005) A History of the World in 6 Glasses. Walker & Company, New York, pp 13–39.
5. Hodges H (1992) Technology in the Ancient World. Barnes & Noble Books, New York, pp 114–115.
6. Joffe AH (1998) Alcohol and Social Complexity in Ancient Western Asia. Curr Anthropol 39:297–322.
7. Katz SH, Voigt MM (1986) Bread and Beer: The Early Use of Cereals in the Human Diet. Expedition 28:23–34.
8. Arnold JP (1911) Origin and History of Beer and Brewing. From Prehistoric Times to the Beginning of Brewing Science and Technology. Alumni Association of the Wahl-Henius Institute of Fermentology, Chicago, pp 85–89.
9. M (2007) Date Beer and Date Wine in Antiquity. Palest Explor Q 139, 1:55–59.
10. Braidwood RJ (1953) Symposium: Did man once live by beer alone? Am Anthro 55:515–526.
11. Hornsey IS (2003) A History of Beer and Brewing. The Royal Society of Chemistry, Cambridge, pp 75–113.
12. SH, Maytag F (1991) Brewing an ancient beer. Archaeology 44:24–33.
13. Damerow P (2012) Sumerian Beer: The Origins of Brewing Technology in Ancient Mesopotamia. Cuneiform Digital Library Journal 2:1–20.
14. Chazan M, Lehner M (1990) An ancient analogy: Pot baked bread in ancient Egypt and Mesopotamia. Paléorient 16(2):21–35.
15. Legras JL, Merdinoglu D, Cornuet JM, Karst F (2007) Bread, beer and wine: Saccharomyces cerevisiae diversity reflects human history. Mol Ecol 16:2091–2102.
16. Murray MA, Boulton N, Heron C (2000) Cereal production and processing. In Nicholson PT, Shaw I (eds) Cabridge University Press, Cambridge, pp 505–536.
17. Hornsey IS (2003) A History of Beer and Brewing. The Royal Society of Chemistry, Cambridge, pp 32–72.
18. Arnold JP (1911) Origin and History of Beer and Brewing. From Prehistoric Times to the Beginning of Brewing Science and Technology. Alumni Association of the Wahl-Henius Institute of Fermentology, Chicago, pp 226–264.
19. Partington, JR (1935) Origins and Development of Applied Chemistry. Longmans, Green and Co., London, pp 195–198.
20. Arnold JP (1911) Origin and History of Beer and Brewing. From Prehistoric Times to the Beginning of Brewing Science and Technology. Alumni Association of the Wahl-Henius Institute of Fermentology, Chicago, pp 108–110.
21. Xenophon (1897) Anabasis. Dakyns HG (trans) In: The Works of Xenophon. Macmillan and Co., London, Book IV.
22. Arnold JP (1911) Origin and History of Beer and Brewing. From Prehistoric Times to the Beginning of Brewing Science and Technology. Alumni Association of the Wahl-Henius Institute of Fermentology, Chicago, pp 358–361.
23. Herodotus (1890) The History of Herodotus. Macaulay GC (trans) Macmillan, London, Book II, verse 77.
24. McGovern PE (2011) The Archaeological and Chemical Hunt for the Origins of Viniculture. In Perard P, Perrot M (eds) Recontres du Clos-Vougeot 2010. Centre Georges Chevrier, Dijon, pp 13–23.

25. Partington, JR (1935) Origins and Development of Applied Chemistry. Longmans, Green and Co., London, p 513.
26. Cavalieri D, McGovern PE, Hartl DL, Mortimer R, Polsinelli M (2003) Evidence for *S. cerevisiae* Fermentation in Ancient Wine. J Mol Evol 57:S226–S232.
27. Fay JC, Benavides JA (2006) Evidence for Domesticated and Wild Populations of Saccharomyces cerevisiae. PLoS Genetics 1:66–71.
28. Civil M (1964) A Hymn to the Beer Goddess and a Drinking Song. In: Biggs RD, Brinkman JA (eds) Studies Presented to A. Leo Oppenheim. The University of Chicago Press, Chicago, pp 67–89.
29. Murray MA, Boulton N, Heron C (2000) Viticulture and wine production. In Nicholson PT, Shaw I (eds) Cabridge University Press, Cambridge, pp 577–608.
30. Michel RH, McGovern PE, Badler VR (1992) Chemical evidence for ancient beer. Nature 360:24.
31. Michel RH, McGovern PE, Badler VR (1993) The First Wine & Beer. Chemical Detection of Ancient Fermented Beverages. Anal Chem 65:408A–413A.
32. Feigl F (1956) Spot tests in organic analysis. Oesper RE (trans), Elsevier Publishing Co., Amsterdam, pp 350–355.
33. Weast RC (ed) (1987) Handbook of Chemistry and Physics, 68th edn. CRC Press, Boca Raton, Florida, p B-207.
34. Frey-Wyssling A (1981) Crystallography of the two hydrates of crystalline calcium oxalate in plants. Amer J Bot 68:130–141.
35. Vallee BL (1998) Alcohol in the Western World. Sci Am June 1:80–85.
36. A (1975) Alcohol, The Delightful Poison. Delacorte Press, New York, pp 4–6.
37. McGovern, PE (2007) Ancient Wine and Beer. In Discovery!: Unearthing the New Treasures of Archaeology. Fagan BM, ed., Thames and Hudson, London.
38. Pritchard JB (1954) The Ancient Near East in Pictures. Relating to the Old Testament. Princeton University Press, Princeton, p 48.
39. Samuel, D (1996) Investigation of Ancient Egyptian Baking and Brewing Methods by Correlative Microscopy. Science 273:488–490.
40. Samuel D (2000) Brewing and baking. In Nicholson PT, Shaw I (eds) Cambridge University Press, Cambridge, pp 577–608.
41. Samuel D (1996) Archaeology of Ancient Egyptian Beer. J Am Soc Brew Chem 54:3–12.
42. Samuel D (1997) Fermentation Technology 3,000 Years Ago – The Archaeology of Ancient Egyptian Beer. SGM Quarterly 24:3–5.
43. Arnold JP (1911) Origin and History of Beer and Brewing. From Prehistoric Times to the Beginning of Brewing Science and Technology. Alumni Association of the Wahl-Henius Institute of Fermentology, Chicago, pp 65–67.
44. Maksoud SA, M. El Hadidl MN, Amer WM (1994) Beer from the early dynasties (3500–3400 cal B.C.) of Upper Egypt, detected by archaeochemical methods. Veget Hist Archaeobot 3:219–224
45. Stillman JM (1924) The Story of Early Chemistry. D. Appleton and Co., New York, pp 161–168.
46. Brock WH (2000) The Chemical Tree. A History of Chemistry. W. W. Norton & Company, New York, p 21.
47. Holmyard EJ (1990) Alchemy. Dover Publications, New York, pp 27–30.
48. Taylor FS (1992) The Alchemists. Barnes & Noble, New York, pp 30–31.
49. Taylor FS (1937) The Origins of Greek Alchemy. Ambix 1:30–47.

Chapter 4
Grape Wine

> The people of the Mediterranean began to emerge from barbarism when they learnt to cultivate the olive and the vine—5th century BCE Greek historian Thucydides [1].

When one generally thinks of wine, it is usually wine produced from the grape that comes to mind, although as previously discussed, other wines most likely predated the traditional grape wine. Grape wine was thought to be originally produced from the wild ancestor of the Eurasian grape, *Vitis vinifera sylvestris* (Fig. 4.1), which was generally limited to the higher elevations of the north [2–5]. Small isolated populations of these wild grapevines can still be found along riverbank forests from the Atlantic coast of Europe to Tajikistan and the western Himalayas. The collection and use of these wild grapes as a food source by Paleolithic hunter-gatherer populations has been well documented at many prehistoric sites across Europe [3].

Although the grapes of the wild ancestor are generally smaller and more sour than cultivated grapes, they are still considered to be suitable for the production of wine [4]. Due to the limited regions of wild grapevine populations, however, early availability of the grape was significantly reduced in comparison to availability of the date palm or cereal grains. Thus, the wide spread and regular use of grape wine is closely associated with the domestication and cultivation of the wild grape (i.e. viticulture or viniculture) [6], resulting in the domesticated Eurasian grape (*Vitis vinifera sativa*) [3, 7] and ultimately the domesticated common grape (*Vitis vinifera vinifera*) (Fig. 4.2) [4, 5, 8].

Wine production is thought to have first appeared in Southwest Asia, probably in the mountainous region between the Black Sea and the Caspian Sea, sometime around 6000 BCE [3, 6, 9], although others have claimed that a "wine culture" was established by at least 7000 BCE [7, 10]. The earliest reported molecular archaeological evidence for large-scale wine production is from the site of Haiji Firus Tepe in the northern Zagros Mountains of Iran dating back to 5400–5000 BCE [2, 4, 5, 11–13]. Reports claim that even earlier chemical evidence, in association with what appear to be remains of domesticated grapes, has been obtained from the early 6th millennium BCE in the Neolithic village of Shulaveris-Gora in the Transcaucasus region of modern Georgia [14]. Domesticated grape

Fig. 4.1 16th Century botanical print of *Vitis vinifera sylvestris* by Ulisse Aldrovandi

Fig. 4.2 Domesticated common grape (Botanical print (dated 1812); photo by Eflon 2010)

pips dating from the beginning of the 6th millennium BCE have also been reported from Chokh in the Dagestan Mountains of the northeastern Caucasus, as well as grape remains dating from the 5th through early 4th millennia BCE from Shomutepe and Shulaveri along the Kura River in Transcaucasia [2, 12]. Other sites, generally dated between about 6000–3000 BCE, have yielded ancient grape remains or wine jars as attested by chemical analyses [10].

Once winemaking had established itself as a viable enterprise in the Neolithic period, wine began to be traded [12]. It has been proposed that viniculture and wine production developed along similar paths in the early civilizations of Egypt and Mesopotamia, resulting from traders introducing royalty and the upper classes to grape wine [2, 15]. As a demand for the drink was built up among the elite in the late 4th millennium BCE, it was extensively traded in pottery jars both overland and along available waterways [2]. During the Early Bronze Age, the production of wines and oils became a focal point for the economy of emerging societies [16]. Ultimately, as these markets developed, the domesticated grapevine was transplanted farther south and into lowland regions, thus allowing local wine production to begin [2, 12]. Such interactions between southern Palestine and the Nile Delta, which account for the prehistoric trade in wine and early historic Egyptian viniculture, are well documented [2]. Wine production then continued to spread and wine became the drink of choice for the Greeks, Romans, and Celts, in part because many areas in Europe were well suited for the growing of the grape [6].

Our current knowledge of the history of wine production comes from a variety of sources, of which artistic, linguistic, and archaeological records play the greatest role. However, unlike the fermented beverages discussed in the previous chapters, a great deal of chemical archeological analyses have been reported on residues related to wine and thus our discussion of wine's history will begin with these studies.

4.1 Chemical Archaeological Studies

Over the last couple decades, considerable advances have been made in the analysis of organic deposits to support the assumed presence of wine [4, 17, 18]. Such residues are typically either preserved contents of ancient vessels or material absorbed in the ceramic matrix of unglazed pottery sherds [4, 17]. Of course, the unambiguous identification of such residues is significantly challenging as a result of potentially low trace levels and the overall chemical complexity of the residues of interest [18]. The main obstacle to identification of these organic samples is the specificity of appropriate *biomarkers*, as many organic compounds are distributed widely in the environment. In addition, complications from biomarker loss or degradation and sample contamination can be complicating issues [4]. Given these limitations, it is critical that rigorous methodology be developed such that the data generated will be sufficiently diagnostic in order to be able to effectively address the questions posed [4, 18].

The earliest chemical analysis of wine-related residues was reported in 1962 utilizing wet chemical spot tests, solubility, and melting point determinations to investigate samples from wine jars recovered from the tomb of Tutankhamun (ca. 1300 BCE) in Thebes (Fig. 4.3) and the Monastery of St. Simeon near Aswan [4].

Although it has been said that precise details of the tests used were not given, the reported detection of potassium carbonate and potassium tartrate led the authors to conclude that the residues were those from wine [4]. Tartaric acid is present in grapes in relatively high concentration [4, 17] (approximately 4,000 ppm [17]) and has since become the favored biomarker for the detection of wine.

Wine residues were then identified in the 1970s from intact and sealed wine vessels of Roman date recovered from the shipwreck of *La Madrague de Giens* (75–60 BCE). One of the vessels was reported to contain both liquid and solid phases, although the sealed contents had been contaminated by sea water [4]. In this case, the samples were reportedly analyzed by gas chromatography (GC) to identify tartaric acid and various phenolic acid derivatives (tannin and anthocyanin degradation products). However, it should be pointed out that while retention times from chromatography techniques can be compared to known standards, retention times are not unique and are thus not conclusive identification of chemical identity in the analysis of unknown samples.

In 1993, chemical analysis was reported of a red deposit from several jars recovered from Godin Tepe (Fig. 4.3), a site of strong Lower Mesopotamian influences in the Zagros Mountains in Iran. The jars date to the Late Uruk period ($\sim$3500–2900 BCE) and based on some contextual evidence, the deposits were proposed to be the remains of wine [7, 19]. The deposits were analyzed by transmission and diffuse-reflectance FT-IR spectroscopy to provide a spectrum that compared favorably to that of the naturally occurring L-(+) isomer of tartaric acid. The presence of tartaric acid was then further supported by a positive Feigl spot test for tartaric acid. As such, the authors felt confident that the jars probably contained wine [19].

The Feigl spot test for tartaric acid [20] (Fig. 4.4) used in this study reacts tartaric acid with [1,1'-binaphthalene]-2,2'-diol (commonly called BINOL) in concentrated sulfuric acid at 85 °C for 30 min to generate a green fluorescent dye [19–23]. While the identity of this green dye is unknown, the spot test is quite sensitive and can detect the presence of tartaric acid down to 10^{-5} g [20]. However, Feigl points out that a number of related acids (Fig. 4.4) also react under these conditions to generate a green dye [20]. As such, the presence of any of these related acids in the samples analyzed could generate a false positive.

The oldest known chemical archeological evidence was then reported in 1996, with the analysis of a pottery jar from the Neolithic village of Jajji Firuz Tepe (Fig. 4.3) in Iran's northern Zagros Mountains [2, 11]. The jar was dated to 5400–5000 BCE and contained an interior yellowish residue that was analyzed by a combination of diffuse-reflectance FT-IR spectroscopy, high pressure liquid chromatography (HPLC) coupled with ultraviolet (UV) spectroscopy, and wet chemical Feigl spot tests [2, 7, 11]. The collective results were consistent with the presence of calcium tartrate and additionally identified a resin of the terebinth tree (*Pistacia atlantica* Desf.). This tree grows abundantly throughout the Middle and Near East, and its resin was widely used as a wine additive in antiquity [2, 11]. Based on the collected data, the authors concluded that the sample was clearly a

4.1 Chemical Archaeological Studies

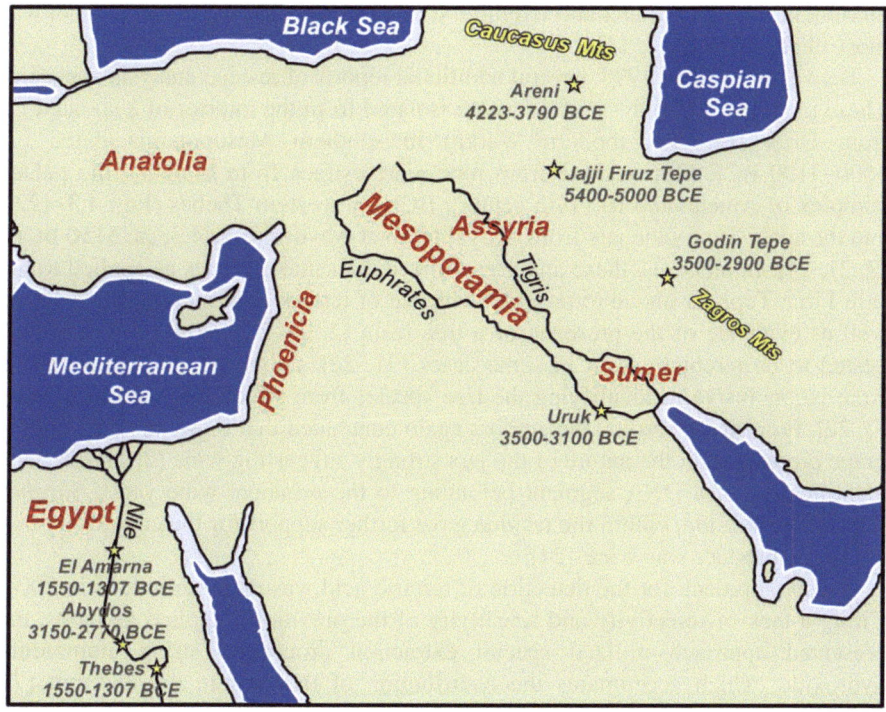

Fig. 4.3 Chemical archaeological sites discussed with associated dating of samples

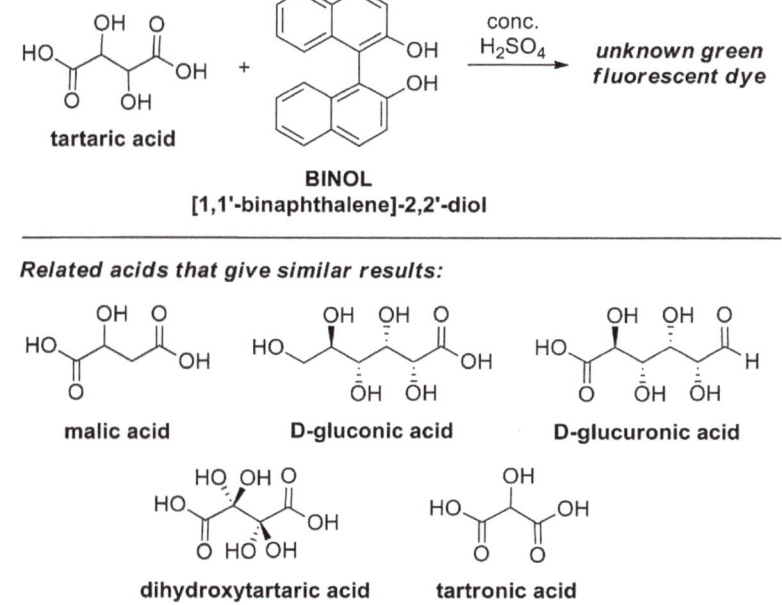

Fig. 4.4 Feigl spot test for tartaric acid with additional acids that give a false positive test

mixture of a grape product and terebinth tree resin, and that the grape product was most likely wine [2, 7, 13].

Between 1996 and 1998, several additional reports of residue analysis appeared. These included the analysis of a residue isolated from the interior of a spouted jar from Uruk (Fig. 4.3, modern Warka) in southern Mesopotamia dated ca. 3500–3100 BCE [23], residue from nine wine ostraca from Malkata, the palace complex of Amenhotep III (14th century BCE) in western Thebes (Fig. 4.3) [22], and the analysis of wine jars from a royal tomb at Abydos (Fig. 4.3, ca. 3150 BCE) [2, 21, 24]. In all cases, these analyses utilized the same methods as applied to the Jajji Firuz Tepe jar above to provide evidence of tartaric acid and tartrate salts, as well as evidence of the presence of a tree resin [2, 21–24]. While this was suggested to be terebinth resin in some cases [21, 22], the chromatographic results were inconclusive in identifying the tree species from which the resin originated [2, 22]. From these results, the authors again concluded that these jars contained a grape product, with the nature of the jars strongly suggesting wine [2, 21–23]. The identification of a DNA segment belonging to the principal wine yeast, *Saccharomyces cerevisiae*, within the residue gave further support for the conclusion that the grape product was wine [24].

A new approach for the detection of tartaric acid was then introduced in 2004. Citing a lack of selectivity and sensitivity of the previously applied methods, the presented approach utilized special extraction procedures using ammonium hydroxide, which accentuates the contribution of the tartrate ion and salt [7]. Liquid chromatography (LC) was then used to separate residue components, which were then directly analyzed using a triple quadrupole tandem mass spectrometer (MS/MS) [25]. Three jars from the British Museum in London and two jars from the Egyptian Museum in Cairo were then analyzed via these new methods. The jars originated from a variety of sites, including Abydos, Nubia, Thebes, and El Amarna, with associated dating ranging from ca. 3100–1300 BCE [25]. Tartaric acid was positively identified in four of the five vessels, although all but one sample required the use of the MS/MS multiple reaction monitoring (MRM) mode, which provides the highest sensitivity. In all cases, the presence of tartaric acid was determined by both LC retention time and MS/MS fragmentation patterns, both of which compared well to the standard analyzed under the same conditions [25].

While the combination of the various methods utilized in many of these analyses provides convincing evidence for the presence of tartaric acid and its salts, these compounds alone do not conclusively prove the former presence of wine. Supporters of its use as an effective biomarker for grapes and wine claim that tartaric acid and its salts occur *"in large amounts only in grapes"* [7, 8, 11, 19] and, as pointed out above, tartaric acid is present in grapes in rather large amounts of $\sim$4,000 ppm. This concentration does vary, however, with both the ripeness and variety of the grape, as well as the geographical latitude in which it is grown [17]. Those that question its use as a reliable biomarker, however, have pointed to multiple plant sources with even higher concentrations of tartaric acid [17]. For example, the fruit of the hawthorn tree (*Crataegus monogyna*) contains about

16,000 ppm of tartaric acid [17, 26], four times that of the grape. In the Near Eastern highlands of Anatolia, hawthorn was used as early as the 2nd millennium BCE, possibly even earlier. Other fruits with tartaric acid content much higher than grapes include tamarind (*Tamarindus indica*, 180,000 ppm), star fruit (*Averrhoa carambola*, 25,000 ppm) and yellow plum (*Spondia mombin*, 15,000 ppm), although none of these are native to the Near East. Even so, it has been pointed out that the amounts of tartaric acid have only been reliably quantified for a limited number of plant species and additional plants may be found to contain high amounts of tartaric acid as more plants are studied [17].

Another complicating factor is the high solubility of tartaric acid in water.[1] As a result, it can readily leach out of buried vessels, as well as potentially leach into adjacent ones, therefore making the presence or absence of tartaric acid in archaeological samples difficult to interpret [17]. This is less so for either its calcium or monopotassium ($KC_4H_5O_6$) salts,[2] although their water solubility is still high enough to result in transfer in archaeological contexts [17]. Dipotassium tartrate ($K_2C_4H_4O_6$), however, exhibits a higher solubility than even the parent acid.[3] Thus tartaric acid and its salts can only serve as reliable biomarkers for the species mentioned above in dry contexts and with favorable conditions for the preservation of organic materials. Lastly, even if the tartaric acid originates from grape, this does not necessitate the presence of wine, but could instead indicate the presence of unfermented grape juice or syrup, unfermented grapes, or even raisins [17].

The same report that introduced the use of LC-MS/MS for the detection of tartaric acid in 2004 also presented an alternate approach for the detection of red wine utilizing the biomarker malvidin (Fig. 4.5) [25]. Malvidin and malvidin-3-glucoside are the major anthocyanins that give grapes and wines their red-purple color and are present in few other plants [17, 25, 27, 28]. Over time, these compounds react with other species in the wine to generate more stable anthocyanin-derived pigments (Fig. 4.5), resulting in a color change to the darker dusky red-brown hue of mature wines [17, 25, 27, 28]. The actual color produced, however, also depends on the pH of the solution [17]. These oligomeric and polymeric malvidin-based pigments exhibit low water solubility and are responsible for the persistent nature of red wine stains. As such, these materials would be expected to be well preserved in archaeological contexts. Unfortunately, the isolation and identification of these pigments has proven difficult due to their large molecular weight, their heterogeneous nature, and the fact that they are present in much lower amounts relative to the parent anthocyanins [17, 27]. Luckily, the treatment of these pigments with strong base releases syringic acid (Fig. 4.5) which can then be detected by analytical methods [17, 25].

---

[1] Water solubility of L-tartaric acid is 139 g/100 mL at 20 °C [29].
[2] Water solubility of calcium tartrate is 0.04 g/100 mL at 10 °C [30]. Water solubility of potassium bitartrate is 0.4 g/100 mL at 10 °C [31].
[3] Water solubility of potassium tartrate is 200 g/100 mL [32].

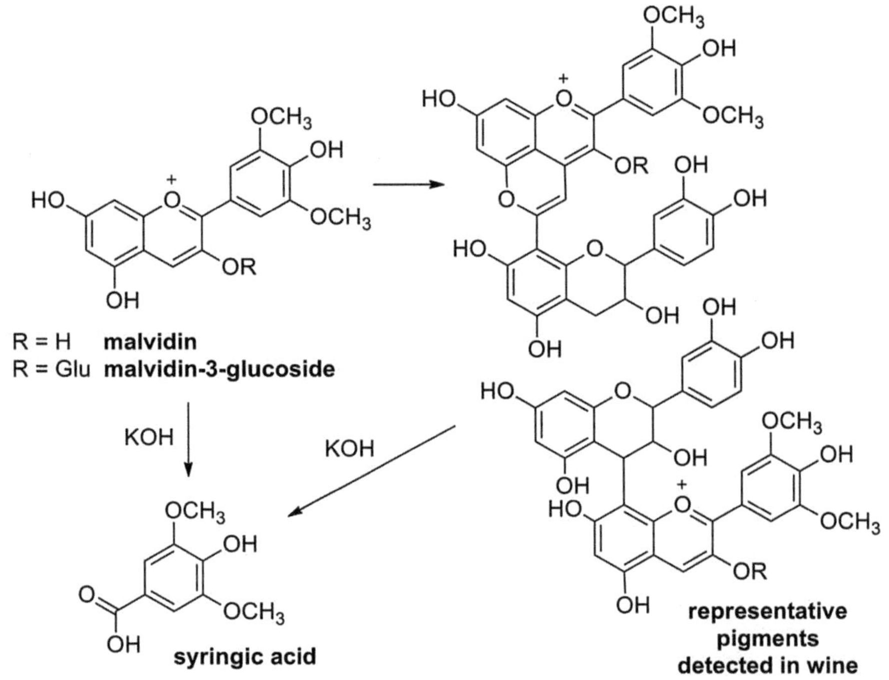

Fig. 4.5 Malvidin and its reactions

To test the viability of malvidin as a biomarker for red wine, the sample from the study above that required the least sensitive methods to detect tartaric acid was then analyzed by LC-MS/MS both before and after treatment with aqueous KOH [25]. Again, using the more sensitive multiple reaction monitoring (MRM) mode of the MS/MS, syringic acid was successfully detected after alkaline treatment. As syringic acid was not detected from analysis before the alkaline treatment, the authors felt confident that the syringic acid detected was produced from the base-mediated cleavage of malvidin [25].

A second study utilizing malvidin as a biomarker was then reported in 2011 to analyze four ancient potsherds from Armenia and Syria [17]. The three Armenian samples were from a site near the village of Areni (Fig. 4.3) dated to 4223–3790 BCE, which based on the presence of desiccated grapes, grape seeds, grape rachises, and grape skins was believed to be a historical grape pressing installation. The Syrian sherd (ca. 2200 BCE) displayed a bright red interior surface resulting from the remains of its former contents and had been suggested to possibly contain red wine [17].

The primary modification to the methodology was the introduction of solid phase extraction using a solvent gradient of increasing polarity to clean the samples and remove any background syringic acid [17]. As syringic acid is present in a number of plant products (including barley, wheat, and wine), as well as soil and other organic environments due to microbiological activity, such sample preparation is critical to remove potential background syringic acid from the polymeric material before the base treatment. The samples were then treated with aqueous KOH to decompose the pigments and analyzed using LC-MS/MS. Positive results for the detection of malvidin were obtained for two of the Armenian potsherds, findings concurrent with the archaeological evidence concerning their context, while the Syrian sherd gave a negative result thus casting doubt on its containing red wine [17].

The authors are careful to point out that malvidin is present in pomegranate, the juice of which was sometimes added to wine. Significant amounts of malvidin are also present in whortleberry (*Vaccinium myrtillus*), red clover (*Trifolium pratense*) and high mallow (*Malva sylvestris*), all of which are used by humans. The amounts of malvidin in plants, however, have only been reliably quantified for a limited number of species, although it is expected to be present in fewer species than tartaric acid [17]. Even so, the detection of malvidin alone cannot be used to reliably specify the presence of grapes and it gives absolutely no information concerning whether or not the analyzed material was a fermented product.

In an attempt to provide more direct evidence of fermentation, authors in 2010 proposed the potential detection of ergosterol by GC-MS as a possible biomarker for yeast [18]. The combination of identification of both yeast and grapes in a residue would then provide more confidence that the residue was from a fermented product. As ergosterol is a sterol specific to the fungal kingdom, its detection in organic pottery residues could provide evidence of the presence of yeasts, but the authors acknowledge that the growth of mold and other fungi on the potsherds could contribute ergosterol and complicate its viability as a useful biomarker [18].

As can be seen by the discussion above, even with the recent advances in the chemical analysis of archaeological samples, the ability to obtain definitive historical data from organic residues is still quite difficult. Some authors have even gone so far as to say it may not ever be possible to get trustworthy results from such samples [4]:

> Ancient organic residues are, quite simply, bad and rather intractable samples; by definition, residues are alteration products, modified by unobservable cultural practices, subjected to poorly understood degradation processes, and often contaminated during burial through to recovery, storage, and even analysis.

4.2 Viniculture

The cultivation and ultimately domestication of *Vitis vinifera sylvestris* eventually led to viniculture (or viticulture as it is sometimes referred). Of course, the question is exactly how early was the Eurasian grapevine domesticated and whether it happened in only one place at a particular time or whether it was domesticated in many different places and times [3, 7, 8, 10]. Two basic divergent hypotheses have been formulated. Some authors favor a restricted origin hypothesis in which domestication took place from a limited wild stock in a single location, followed by the transplantation of those cultivars to other regions [3, 7, 8]. This hypothesis has been referred to as the 'Noah Hypothesis' [5, 7, 8], as one of the biblical patriarch's first acts following the great flood was to plant a vineyard and then make wine [33]:

> Noah, a man of the soil, proceeded to plant a vineyard. When he drank some of its wine, he became drunk...

Others envision a multiple-origin hypothesis in which domestication could have involved a large number of originating stock evolving over an extended time period and along the entire distribution range of the wild antecedent species [3].

In an attempt to address these two hypotheses, researchers analyzed chlorotype variation and distribution in 1201 samples of *sylvestris* (wild type) and *sativa* (domesticated) genotypes from across the regions of the species' distribution in order to study their genetic relationships. The results support the existence of a relevant genetic contribution of eastern and western *sylvestris* population groups to the genetic makeup of current grapevine cultivars and could suggest the existence of at least two origins for the cultivated grapevine, one in the Near East and another in the western Mediterranean region, the latter of which could have been the origin to many of the current Western European cultivars [3]. However, the authors are careful to point out that it is unclear whether this second origin represents an independent domestication event or developed as a consequence of the east to west transmission of the cultivated vine. As such, further archaeological research is still needed to address this question [3].

Despite some disagreement about the exact location (or locations) of the first domestication event [4], it is clear that by the 4th millennium BCE, the Mediterranean crops of grapes, olives, dates, and figs were all being used in the Levant [16]. The domestication of the grapevine has been linked to the production of wine, which required both suitable access to grapes as well as storage containers made of pottery to preserve the wine. Such developments are generally not thought to have occurred prior to the Neolithic period (ca. 10000–4000 BCE) [3, 8, 10].

In terms of the geographical origins of viniculture, one must look to where *Vitis vinifera sylvestris* thrives, typically further north and at higher elevations. Neolithic communities in upland regions of the northern Zagros Mountains, the eastern Taurus Mountains, and the Caucasus Mountains were well established from an early time and, based on current archaeological and historical knowledge, are probably the best candidates for early winemaking and viniculture [2, 3, 5, 7, 10, 12, 34]. In fact, many point to the Transcaucasia region, where the greatest genetic diversity of the grape is found, and ca. 6000 BCE for the beginnings of viniculture (Fig. 4.6) [3–6]. The belief that the Transcaucasia region may be the original site of wine production is also supported by the theory that the hypothetical proto-Indo-European root for the word 'wine' (*woi-no or *wei-no) had its origin in eastern Turkey or Transcaucasia [5]. Unfortunately, few sites have been excavated within this ethnically diverse and politically divided region, let alone published in a Western language [2, 8]. However, archaeological research has found grape pips of the domesticated Eurasian grape from Chokh in the Dagestan Mountains of the

4.2 Viniculture

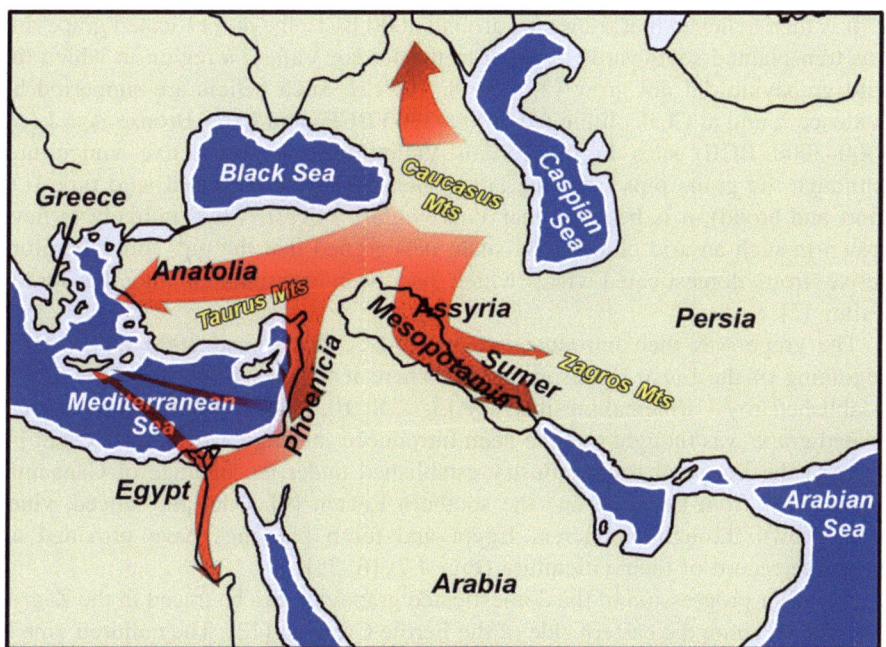

Fig. 4.6 Proposed path of the transmission of viniculture based upon archaeological findings

northeast Caucasus, dating to the beginning of the 6th millennium BCE, as well as from Shomutepe and Shulaveri along the Kura River in Transcaucasia, dating to the 6th through early 4th millennium BCE [2, 8, 12].

While the Transcaucasia region is the favored site for the origin of viniculture, the details concerning the introduction of cultivated grapevines into the southern regions remains a topic of debate. It is generally believed that winemaking moved from its source to the cultures of the Sumerians, Assyrians, Babylonians, and Hittites [5] and the appearance of the cultivated grapevine into Mesopotamia has been dated as early as 6000 BCE [35, 36]. Others, however, have pointed to later dates for this event.

Potential evidence to support suggested timelines include archaeobotanical samples from Kurban Hüyük, in Anatolia, which have indicated that grapes were present there in increasing quantities in 4th- and 3rd-millennium BCE [16]. In terms of Mesopotamia, it has been reported that grapes were likely introduced to Iran and the Mesopotamian lowlands from the north, but suggest that this occurred in the early 4th to late 3rd millennium BCE, as the Sumerian sign for grapes does not appear until then [16]. This timeline, however, is complicated by another sign interpreted as wine, which appeared as early as the mid-4th millennium BCE. However, it is only in the second half of the 3rd millennium and thereafter that references to grapes, raisins, and wine become increasingly frequent in cuneiform sources [16].

It is then believed that sometime around 4000 BCE, the domesticated grapevine was transplanted southwards to reach to the Jordan Valley, a region in which the wild grapevine did not grow [2, 3, 7, 8, 10, 12]. Such beliefs are supported by evidence found at Chalcolithic (ca. 4000–3300 BCE) and Early Bronze Age I (ca. 3300–3000 BCE) sites in the Jordan Valley that suggest active viniculture. Although the grape pips from the latter sites are of the proposed wild type (i.e. short and broad), it is believed that *Vitis vinifera sylvestris* was unlikely to have grown in such an arid climate. As such, it is argued that the pips must therefore derive from domesticated vines which had been transplanted into the Jordan Valley [2].

The grape was then introduced to the Nile Delta around 3000 BCE, at the beginning of the Early Dynastic period, where it formed the basis of the newly established royal winemaking industry [3, 7, 8, 10, 12, 13, 15, 16]. The domesticated grape was thought to have been introduced into Egypt from the Levant [4, 7], with the Egyptian royal industry established under the tutelage of Canaanite winemakers from Lebanon and the southern Levant [9]. Once introduced, vines were grown throughout ancient Egypt and tomb paintings have provided an extensive record of their viticulture (Fig. 4.7) [6, 25].

A similar progression of the domesticated grapevine can be traced in the Zagros Mountains, along the eastern side of the Fertile Crescent [12]. The cultured vine is believed to have spread to the central and southern Zagros Mountains, bordering Mesopotamia on the east, by ca. 3000 BCE [3, 10]. Grape pips and even grapevine wood have been identified from the late 4th–3rd millennium BCE site of Tepe Malyan in the southern Zagros, indicating that the domesticated plant had already been transplanted to the southern Zagros by at least the mid-3rd millennium BCE [12].

Viniculture also emerged in Asia Minor and Greece sometime during the 4th–3rd millennium BCE. It was during this time period that grape cultivation is thought to have moved from being just an aspect of local consumption to an important component of local economies and trade. Western expansion of the wine culture is documented in Crete, c. 2200 BCE [3, 15], and later on the coasts of the Italian and Iberian Peninsulas (c. 800 BCE) [3].

Although it is thought by some that viniculture was introduced into Italy as early as 1000 BCE, it has been stated that wine in Rome was still a rather scarce commodity in the 4th century BCE [34]. It is thought that Italian viniculture came to play an important role by the beginning of the 2nd century BCE, and became widespread in Italy after 150 BCE, although the quality of the wine was far from anything that would have given it a world reputation. Nevertheless it is thought that local wines now satisfied internal demand, as the importation of Greek wine was so limited by this time that it was sparingly used even by the wealthy [34]. Once introduced into Greece, Italy, and Gaul, wine became a beverage used by all levels of society [6]. A general timeline of the cultivation of the grapevine is summarized in Table 4.1.

The initial cultivation and domestication process appears to have involved the favoring and selection of hermaphroditic members of the wild species over either

4.2 Viniculture

Fig. 4.7 Egyptian depiction of cultivated grapevines [37]

Table 4.1 General timeline of vine cultivation [3, 7, 10, 12, 15, 16, 35, 36, 38]

Time period	Region
6000 BCE	Caucasus
6000–4500 BCE	Mesopotamia
4000 BCE	Jordan Valley
3000 BCE	Egypt, Phoenicia
2500–2000 BCE	Greece, Crete
1000 BCE	Italy, Sicily, North Africa
500 BCE	Spain, Portugal, Southern France
100 BCE	China, Northern India
100 CE	Northern Europe
1500 CE	North America

the barren male vines or female vines which were dependent on having a nearby male to pollinate [2, 3, 11]. As such hermaphrodite genotypes include the male (stamens) and female (pistil) parts located on the same vine, these plants have the ability to pollinate themselves and can produce more fruit on a predictable basis [2, 7, 10]. Over time, these hermaphroditic vines were able to sire offspring that was consistently hermaphroditic itself [2] and this self-fertilizing plant could then be selected for larger, juicer and tastier fruit and fewer seeds [3, 7, 10]. The final domestication process then involved the development of techniques for their vegetative propagation, or cloning, by transporting branches, buds, or rootings [3, 4, 7, 10]. Such cloning methods eliminated the need for sexual reproduction via seeds and resulted in consistent and reproducible progeny. As a result, desirable qualities, such as abundant fruit production, high juice content, flavor, and color could all be selected and reproduced [4, 7].

4.3 Wine Production

In both Mesopotamia and Egypt, vineyards were thought to be limited to the property of the rulers and grape wine was not typically available to the common man. It has even been suggested that grape wine was reserved entirely for the gods [39] and its use as an offering to the gods is known, with both wine and grapes being regularly placed in tombs [40]. At least initially, the consumption of wine was restricted to the upper classes due its extreme cost [15, 40]. This was due to the fact that during the introduction of the beverage, the cultivation of the grapevine in the south had not yet occurred and all production was still carried out in the mountainous regions to the northeast and then exported to the south [15]. As such, the cost of transporting wine down from the mountains to the plains of Mesopotamia made it at least ten times more expensive than beer, and was initially regarded as an exotic foreign drink [15]. Textual sources indicate that by the 2nd millennium BCE, the northern Mesopotamian city of Mari had become a major importer and distributor of wines, a conclusion supported by archaeological evidence [16].

Even once wine production was established in the southern regions the import and export of wine still played a significant role. For example, it has been reported that while most wine produced in the Levant appears to have been intended for internal distribution and consumption, an unknown amount of the total production was exported to Egypt [16], which imported such wines primarily as a drink of the aristocratic classes [4, 7]. This wine was imported primarily from Syria and Palestine, but also from Greece in the later periods (ca. 6th century BCE) [7, 40]. Herodotos documents that this wine was imported twice a year from Greece and Phoenicia in earthen jars [40]. Once Egypt began producing its own wines, the dependency on Levantine wines diminished, resulting in a corresponding change in the pattern of Levantine wine consumption in Egypt [16]. The best wines produced in Egypt have been said to be those that came from the Nile River Delta and the Western oasis [25].

Wine trade had enormous economic significance during the Roman era. The wine amphora became one of the most common ceramics of the time [6]. As shown in Fig. 4.8, the amphora was a type of container of a characteristic shape and size, which were used in vast numbers for the transport and storage of various products, both liquid and dry, but mostly for wine. It has been reported that an estimated 40 million amphorae of wine were imported into Celtic territories during a single century at the end of the Iron Age, corresponding to 2.65 million gallons of wine per year [6].

The actual processing of grapes into wine is not necessarily that complex, as illustrated by textual and pictorial evidence of wine production from the ancient world [19]. While the processes utilized in Egypt are typically the most detailed, these methods of wine production are thought to be representative of other regions and cultures as well. Overall, available sources seem to indicate that the processes

4.3 Wine Production

Fig. 4.8 A typical Roman wine amphora

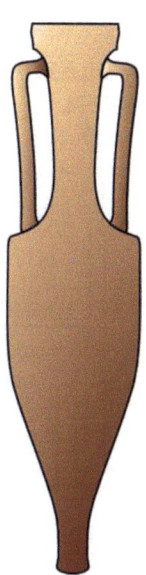

for the production of grape wine used in antiquity generally follow the same processes known today [39].

The production of wine began with the harvesting of the grapes [4, 39], with the collected grapes placed into baskets for transportation [4]. Such baskets of grapes have sometimes been shown covered with palm leaves or vines, which have been theorized to be forms to protect the collected fruit from the sun [4]. The baskets of harvested grapes were then emptied into large vats used to tread the juice from the grapes [4].

The traditional method used to crush the grapes throughout the Mediterranean and Near East was to tread them underfoot in crushing vats [4, 19, 39, 40]. Pictorial evidence from Egyptian tombs clearly illustrate that this method was also employed in Egypt (Fig. 4.9) [4]. Such methods are gentler than modern wine presses and are considered the most effective way to release the grape juice without also crushing large quantities of the seeds and stems [4]. Crushed seeds and stems can add unwanted tannins, astringency, and color to the resulting grape juice [4].

In Egyptian scenes of the 3rd millennium BCE, the treading vat is shown as very shallow and of indeterminable shape. At either end of the vat are upright poles, sometimes forked, which support a crossbar. This crossbar is then utilized by those treading the grapes to increase balance and more efficient treading of the grapes [4]. Later scenes also show a type of vat that appears to be rounded and much deeper than the earlier forms, but retains the upright poles and crossbars. Frequently, these later vats also utilize a spout from which the extracted juice pours into a waiting container [4].

A final type of treading vat depicted in some scenes is much more elaborate and often appears on a raised platform, with steps leading up to it. In addition to the

Fig. 4.9 Egyptian scene from a Theban tomb showing the treading of grapes [29]

upright poles and crossbars pictured in the other forms, these vats are sometimes shown with a full roof from which support ropes hang down for those treading the grapes (Fig. 4.9). Vats of this final form also contain a spout at the base for collection of the juice [4].

It has been estimated that only about two-thirds of the juice can be extracted from the grape by foot treading [4]. For more efficient collection of the juice, many tomb scenes show that the mash resulting from the treading vats (i.e. crushed skins, stalks and seeds) could then be placed into a bag press (Fig. 4.10). Such presses could then be used to squeeze additional juice remaining in the mash residue [4, 19, 40].

The earliest such bag presses consisted of a simple bag or sack with poles attached at each end (Fig. 4.10a). The poles could then be twisted to squeeze the bag until any remaining juice had been extracted. It has been proposed that the bag material might have been made of linen or possibly consisted of some type of basketry [4]. Advances of the 2nd millennium BCE included first attaching one end of the bag to a fixed post, requiring only a single pole to twist the bag (Fig. 4.10b). Finally, a more advanced bag press was introduced sometime after the 16th century BCE, consisting of one or both ends of the bag attached to poles outside of a frame (Fig. 4.10c, d). The addition of the frame allowed the bag to be fixed securely in place, thus allowing the poles at either end to be twisted using maximum leverage [4].

The isolated juice extracted from the grape that has been exposed to yeasts is referred to as grape *must* or the *must* of wine [4, 19]. The necessary yeasts are present in the "bloom" of the grape skins and thus mix with the juice during isolation via treading and pressing [19]. The "bloom" is the white, powdery coating visible on some grapes found primarily near the grape stalk [4]. As previously discussed for other fermentation processes, the primary yeast involved in

4.3 Wine Production

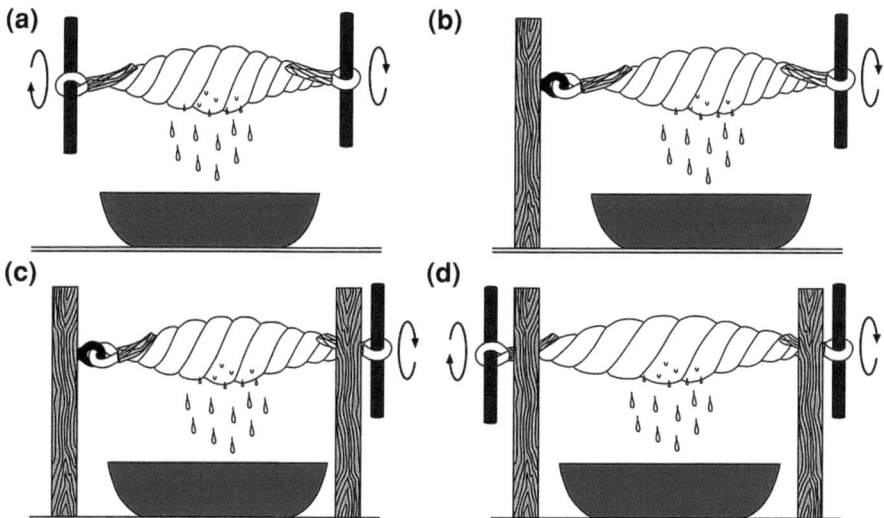

Fig. 4.10 Evolution of the bag press

the fermentation of the grape juice is *Saccharomyces cerevisiae*, although a variety of wild yeasts of the genera *Saccharomyces* and *Candida* have been reportedly found in the grape "bloom" [19]. However, it has been pointed out by some authors that contrary to popular belief, *Saccharomyces cerevisiae* occurs at extremely low populations on healthy, undamaged grapes and is rarely isolated from intact berries and vineyard soils. In fact, it is yeasts of the genera *Kloeckera* and *Hanseniaspora* that are the predominant species on the surface of grapes, accounting for roughly 50–75 % of the total yeast population [35]. Nevertheless, *Saccharomyces cerevisiae* is still by far the most dominant yeast species colonizing surfaces in wineries, which is said to demonstrate the selective effects of grape juice and wine as growth substrates [35]. It wasn't until the 19th century that wine must was produced from the addition of pure yeast cultures to sterile grape juice [6, 35].

Different grades of must are produced during the various stages of isolation and extraction. 'Free-run' must is obtained prior to treading and pressing and consists of the juice pressed out of the grape simply by their own weight. As one might imagine, comparatively little of this must is produced and if collected and fermented without the addition of other juice, produces a pure, sugar-rich, and long-lived wine [4]. Most of the juice used for wine, however, is obtained during treading, which is known as 'first run' must. The additional juice obtained from pressing is then referred to as 'second run' must. The level of impurities increases with the various grades of must, as does the potential to generate either acetic or lactic acid [4]. It is unclear if each type of must was fermented separately or if they were eventually mixed together.

After isolation, the grape must is transferred to large, wide-mouth pottery jars and allowed to ferment [19, 40]. The grape contains both glucose and fructose, which make up about 20 % of the fruit [6, 35]. The fermentation process then converts these sugars to ethanol until the alcohol content reaches between 13 and 16 %, at which point the yeast is spent and fermentation stops [4]. At this point, the fermented material was either strained and transferred to secondary sealed containers [4, 39] or the original fermentation container was stopped and sealed [4, 19].

Wine was commonly stored in large amphorae (Fig. 4.8), pitched outside and closed with a stopper [37], and the wine cooled by fanning the jars [40]. It has been suggested that the insides of these jars were coated with resin or bitumen to render them impermeable. Such resins could also help preserve the wine, as will be further discussed below [4]. The stoppers utilized generally consisted of a variety of materials, including reeds, straw, pottery, wood, or clay [4, 37]. After the amphorae were stoppered, they were completed with either hand-made or molded seals [4, 37, 40]. Those made by hand simply consisted of placing an amount of dampened mud on the mouth of the jar, shaping it by hand, and allowing it to dry [4, 37]. Molded seals were more elaborate and were made by either filling an open-ended mold with mud and forcing it onto the neck of the amphorae in order to enclose it, or covering the neck with mud and then applying the mold [4]. The rapid sealing of the wine containers was important as the continued availability of oxygen after fermentation was finished could result in the growth of bacteria responsible for the conversion of ethanol to acetic acid [4, 19].

Alcohol-soluble plant resins were sometimes added to wine, including pine, cedar, frankincense, myrrh, and terebinth [2, 7, 8, 10, 11, 40]. Pliny the Elder describes the addition of pitch and resins to wines in his *Naturalis Historia* [41]:

> The method used for seasoning wines is to sprinkle pitch in the must during the first fermentation, which never lasts beyond nine days at the most, so that a bouquet is imparted to the wine, with, in some degree, its own peculiar piquancy of flavour. It is generally considered, that this is done most effectually by the use of raw flower of resin, which imparts a considerable degree of briskness to wine: while, on the other hand, it is thought that crapula itself, if mixed, tends to mitigate the harshness of the wine and subdue its asperity, and when the wine is thin and flat, to give it additional strength and body. It is in Liguria more particularly, and the districts in the vicinity of the Padus, that the utility is recognized of mixing crapula with the must, in doing which the following rule is adopted: with wines of a strong and generous nature they mix a larger quantity, while with those that are poor and thin they use it more sparingly. There are some who would have the wine seasoned with both crapula and flower of resin at the same time. Pitch too, when used for this purpose, has much the same properties as must when so employed.

In addition to flavoring the wine, these resins would inhibit the growth of acetic acid bacteria (*Acetobacter aceti*, *Acetobacter pasteurianus*, and *Gluconobacter oxydans*) that convert wine to vinegar [2, 7, 10, 35]. This use of pitch and resins to preserve wine is detailed by Columella in his *De re rustica* [42]:

> When we shall have, in this manner, prepared the pitch, and have a mind to preserve our wines therewith, when they have now twice left off fermenting, we must put two cyathi of

the foresaid pitch into forty-eight sextarii of must, in this manner: We must take two sextarii of must out of that quantity we are going to preserve; then, from these two sextarii, we must, by little and little, pour the must into the two cyathi of pitch, and work it with our hand as it were honey and water, that it may the more easily mix with the must: but, when the who two sextarii of must are mingled with the pitch, and make, as it were, an unity of substance, then it will be proper to pour them into the vessel from whence we took them, and to stir it about with a wooden ladle, that the medicament may be throughly mixed with it.

The practice of adding resins to wine is thought to have been introduced into Egypt during the Ptolemaic period (323–30 BCE) [4]. Others, however, assert that the application of terebinth resin in wine dates back to the 6th millennium BCE [8].

With few exceptions, both white and black grapes produce a white juice, although most modern wine is made primarily from white grapes. Contrary to common perceptions, wine color is generally not related to the grape used, but is determined by how long the grape skins and seeds are left in the juice to be isolated [4]. White wine is produced from the initial clear juice and red wine from juice containing the skins and seeds for a longer period [4, 6]. While there is little explicit evidence of the color of Egyptian wines, artistic and textual records both suggest that red wine predominated [4]. However, it has been reported that at least six kinds of wine are mentioned in Egyptian sources, including white, red, black, spiced, and wine of the Delta [40]. In a similar manner, Bronze Age texts from Mesopotamia only mention the color red in relation to wine [4]. The reference to spiced wines is consistent with the fact that grape wines were sometimes flavored with herbs [39].

References

1. Johnson H (1989). Vintage: The Story of Wine. Simon and Schuster, New York, p 35.
2. McGovern PE, Hartung, U, Badler VR, Glusker DL, Exner LJ (1997) The beginnings of winemaking and viniculture in the ancient Near East and Egypt. Expedition 39:3–21.
3. Arroyo-García R, Ruiz-García L, Bolling L, Ocete R, López MA, Arnold C, Ergul A, Söylemezoğlu G, Uzun HI, Cabello F, Ibanez J, Aradhya MK, Atanassov A, Atanassov I, Balint S, Cenis JL, Costantini L, Gorislavets S, Grando MS, Klein BY, McGovern PE, Merdinoglu D, Pejic I, Pelsy F, Primikirios N, Risovannaya V, Roubelakis-Angelakis KA, Snoussi H, Sotiri P, Tamhankar S, This P, Troshin L, Malpica JM, Lefort F, Martinez-Zapater JM (2006) Multiple origins of cultivated grapevine (*Vitis vinifera* L. ssp. *sativa*) based on chloroplast DNA polymorphisms. Mol Ecol 15:3707–3714.
4. Murray MA, Boulton N, Heron C (2000) Viticulture and wine production. In Nicholson PT, Shaw I (eds) Cabridge University Press, Cambridge, pp 577–608.
5. McGovern PE (2004) Southeastern Turkey: Homeland of Winemaking and Viticulture? ARIT Newsletter 38:10–11.
6. Lambert JB (1997) Traces of the Past. Unraveling the Secrets of Archaeology through Chemistry. Addison-Wesley, Reading, MA, pp 136–137.
7. McGovern PE (2012) The Archaeological and Chemical Hunt for the Origins of Viniculture in the Near East and Etruria. In Archeologia della Vite e del Vino in Toscano e nel Lazio: Dalle

tecniche dell'indagine archeologica alle prospettive della biologia molecolare. Ciacci A, Rendini P, Zifferero A (eds). Edizioni all'Insegna del Giglio: Borgo S. Lorenzo, pp 141–152.
8. McGovern P (2002) Wine and the Vine: New Archaeological and Chemical Perspectives on Its Earliest History. In Bacchus to the Future. C. Cullen C, Pickering G, Phillips R (eds) Brock University: St. Catherines (Ontario, Canada), pp 565–592.
9. Vallee BL (1998) Alcohol in the Western World. Sci Am June 1:80–85.
10. McGovern PE (2011) The Archaeological and Chemical Hunt for the Origins of Viniculture. In Perard P, Perrot M (eds) Recontres du Clos-Vougeot 2010. Centre Georges Chevrier, Dijon, pp 13–23.
11. McGovern PE, Glusker DL, Exner LJ, Volgt MM (1996) Neolithic resonated wine. Nature 381:480–481.
12. McGovern PE (1998) Wine's Prehistory. Archaeology 51:32–34.
13. McGovern PE (2007) Ancient Wine and Beer. In Discovery!: Unearthing the New Treasures of Archaeology. Fagan BM (ed). Thames and Hudson: London, pp 232–233.
14. Cavalieri D, McGovern PE, Hartl DL, Mortimer R, Polsinelli M (2003) Evidence for S. cerevisiae Fermentation in Ancient Wine. J Mol Evol 57:S226–S232.
15. Standage, T (2005) A History of the World in 6 Glasses. Walker & Company, New York, pp 43–56.
16. Joffe AH (1998) Alcohol and Social Complexity in Ancient Western Asia. Curr Anthropol 39:297–322.
17. Barnard H, Dooley AN, Areshian G, Gasparyan B, Faull KF (2011) Chemical evidence for wine production around 4000 BCE in the Late Chalcolithic Near Eastern highlands. J Archaeol Sci 38:977–984.
18. Isaksson S, Karlsson C, Eriksson T (2010) Ergosterol (5,7,22-ergostatrien-3β-ol) as a potential biomarker for alcohol fermentation in lipid residues from prehistoric pottery. J Archaeol Sci 37:3263–3268.
19. Michel RH, McGovern PE, Badler VR (1993) The First Wine & Beer. Chemical Detection of Ancient Fermented Beverages. Anal Chem 65:408A–413A.
20. Feigl F (1956) Spot tests in organic analysis. Oesper RE (trans), Elsevier Publishing Co., Amsterdam, p 358.
21. McGovern PE (1998) Wine for Eternity. Archaeology 51:28–32.
22. McGovern PE (1997) Wine of Egypts Golden Age: An Archaeochemical Perspective. J Egypt Archaeol 83:69–108.
23. Badler VR, McGovern PE, Glusker DL (1996) Chemical evidence for a wine residue from Warka (Uruk) inside a Later Uruk Period spouted jar. Baghdader Mitteilungen 27:39–43.
24. McGovern PE, Mirzoian A, Hall GR (2009) Proc Natl Acad Sci USA 106:7361–7366.
25. Guasch-Jané MR, Ibern-Gómez M, Andrés-Lacueva C, Jáuregui O, Lamuela-Raventós RM (2004) Liquid chromatography with mass spectrometry in tandem mode applied for the identification of wine makers in residues from ancient Egyptian vessels. Anal Chem 76:1672–1677.
26. McGovern PE, Zhang J, Tang J, Zhang Z, Hall GR, Moreau RA, Nunez A, Butrym ED, Richards MP, Wang C, Cheng G, Zhao Z, Wang C (2004) Fermented beverages of pre- and proto-historic China. Proc Natl Acad Sci USA 101:17593–17598.
27. Mateusa N, de Pascual-Teresab S, Rivas-Gonzalo JC, Santos-Buelgab C, de Freitasa V (2002) Structural diversity of anthocyanin-derived pigments in port wines. Food Chem 76:335–342.
28. Remy S, Fulcrand H, Labarbe B, Cheynier V, Moutounet M (2000) First confirmation in red wine of products resulting from direct anthocyanin–tannin reactions. J Sci Food Agric 80:745–751.
29. Windholz M (ed) (1983) The Merck Index, Tenth Edition. Merck & Co., Inc., Rahway, NJ, p 1303.
30. Windholz M (ed) (1983) The Merck Index, Tenth Edition. Merck & Co., Inc., Rahway, NJ, p 235.

References

31. Windholz M (ed) (1983) The Merck Index, Tenth Edition. Merck & Co., Inc., Rahway, NJ, p 1099.
32. Windholz M (ed) (1983) The Merck Index, Tenth Edition. Merck & Co., Inc., Rahway, NJ, p 1105.
33. Barker K (ed) (1995) The NIV Study Bible, 10th ann ed, Genesis, 9:20–21.
34. Jellinek EM (1976) Drinkers and alcoholics in Ancient Rome. J Stud Alcohol 37:1718–1741.
35. Pretorius IS (2000) Tailoring wine yeast for the new millennium: novel approaches to the ancient art of winemaking. Yeast 16:675–729.
36. Crane E (1999) The World History of Beekeeping and Honey Hunting. Routledge, New York, p 513.
37. Arnold JP (1911) Origin and History of Beer and Brewing. From Prehistoric Times to the Beginning of Brewing Science and Technology. Alumni Association of the Wahl-Henius Institute of Fermentology, Chicago, pp 66–67.
38. Legras JL, Merdinoglu D, Cornuet JM, Karst F (2007) Bread, beer and wine: Saccharomyces cerevisiae diversity reflects human history. Mol Ecol 16:2091–2102.
39. Hodges H (1992) Technology in the Ancient World. Barnes & Noble Books, New York, pp 114–117.
40. Partington, JR (1935) Origins and Development of Applied Chemistry. Longmans, Green and Co., London, p 198–199.
41. Pliny the Elder (1855) The Natural History. Bostock J, Riley HT (trans) Taylor and Francis, London, Book XIV, Chapter 25.
42. Columella LJM (1745) L. Junius Moderatus Columella of Husbandry in Twelve Books and his Book concerning Trees. Millar A (trans) London, UK, Book XII, p 532.

Chapter 5
Fermented Milk

In addition to the more traditional wines from the fermentation of fruits, and beer from the fermentation of grains, some cultures also developed fermented alcoholic beverages from the milk of various animals. Of course, this required the domestication of animals in order to provide a regular supply of milk. As such, fermented milk products were generally developed at later dates then either beer or wine.

Sheep and goats were the first livestock species to be domesticated [1, 2], this occurring about 9000–8000 BCE [1]. It is believed that multiple domestication events, as inferred by multiple mitochondrial lineages, gave rise to domestic sheep and similarly other domestic species [2]. Initially, sheep were reared mainly for meat, but dairying developed relatively quickly after domestication [3] and the consumption of animal milk is thought to date to the mid-6th millennium BCE or earlier [4]. The availability of a regular supply of milk ultimately led to the systematic production of alcoholic drinks such as kefir and kumis via the fermentation of animal milk [5].

Archaeological evidence of various ancient civilizations suggests that the fermentation of milk has been known for millennia [6]. Although the origins of such alcoholic milk drinks are unknown, it is thought that such fermented milk products likely originated in the Middle East and the Balkans [7]. It has been reported that a number of central and northern Asiatic pastoralist societies (e.g. Scythians, Sarmatians and Huns) were familiar with soured milk drinks from the 1st millennium BCE [4, 5]. Others, however, have placed the beginning of the manufacture of fermented milks back to the 2nd millennium BCE [6].

Kefir and kumis are the most common examples of fermented alcoholic milk drinks, and are produced via the use of certain strains of lactic acid bacteria and yeasts [8]. The specific microflora involved, however, is fairly complex and not always constant, with numerous species of lactic acid bacteria, yeasts, and molds potentially involved [7]. Such alcoholic drinks produced via yeast-lactic fermentation are characterized by a white or yellowish color, a balanced slightly yeast-like aroma, a slightly tart and refreshing taste, and a thick consistency [8].

5.1 Kefir

Kefir[1] is popularly believed to have originated centuries ago among the shepherds of the Caucasus mountain region and is made mainly in the Caucasus region, the Balkans and the Middle East [6, 9–12]. It is produced by adding kefir grains to animal milk and letting it ferment for days in clay pots, wooden buckets, or animal-skin bags [6, 9, 10]. Kefir is produced mainly from cow milk, but has also been made from goat, buffalo, or sheep milk [6, 8, 10]. This fermentation process results in the production of a sour, carbonated beverage with an alcohol content of ca. 1.5 %. Kefir is characterized by a slightly creamy texture and a distinct, yeasty taste [6].

The traditional method of kefir production starts with the addition of a 0.5–10 % inoculum of kefir grains to heat-treated milk. The predominant process during the first 12–24 h is thought to be lactic acid fermentation, after which the kefir grains are strained and removed to produce a weak kefir. This kefir is then bottled and stored for one to three days, during which time alcoholic fermentation occurs resulting in a stronger kefir [6, 9].

The key to kefir production is the kefir grains added to start the fermentation process. The origin of kefir grains is unknown [9], but it has been suggested that during the continuous use of the same containers to make kefir, their walls became covered with colonies of microorganisms that resemble boiled rice (Fig. 5.1) [6]. Kefir grains are gelatinous, whitish or yellowish, irregular granules which range in size from 1–6 mm. Active kefir grains float on the milk's surface and contain a combination of lactobacilli, streptococci, yeasts, and acetic acid bacteria, all held together in protocooperative or potentially symbiotic relationships [6, 12]. It has also been reported that the surfaces of kefir grains are sometimes covered with a white mold, *Geotrichum candidum*, but this is thought to not affect the grains' performance [6]. Others have also suggested that the microbial flora of the kefir grains seem to differ according to the place of the origin [10].

The microbial population of kefir grains are embedded in a polysaccharide, given the name *kefiran*, that is a capsular material of *Lactobacillus brevis* [6, 11]. Thus, unlike other fermented milk drinks, kefir is not produced by the activity of an evenly distributed microflora, but of the microbes in kefir grains that can be recovered readily after fermentation [10]. Such microbes include the bacteria species *Lactobacillus brevis*, *Lactobacillus fermentum*, and *Lactobacillus kefir* [6, 9]. These lactobacilli ferment hexoses such as glucose to lactic acid, ethanol, and CO_2 [6].

The alcohol content in kefir is produced via the fermentation of the milk sugar lactose, a disaccharide derived from galactose and glucose. The lactose content in cow's milk is roughly 4.9 % by weight [13]. In 1889, while studying fermentation of lactose by the kefir yeast *Saccharomyces kefyr* (now known as *Kluyveromyces*

---

[1] Also known as *kifir* or *khiafar* [9].

Fig. 5.1 Kefir grains (Photo by A. Kniesel, 2005)

marxianus), Martinus Beijerinck (1851–1931)[2] detected a new enzyme which he named lactase (now known as β-galactosidase) [6, 11, 14]. Beijerinck concluded that lactose fermentation by the yeast was preceded by hydrolysis of the disaccharide, a process catalyzed by the action of lactase [14]. As shown in Fig. 5.2, this catalytic hydrolysis thus produces equal equivalents of glucose and galactose.

The monosaccharides resulting from the β-galactosidase-catalyzed hydrolysis of lactose can then undergo lactic acid and alcoholic fermentation as catalyzed by lactobacilli, yeast, or other microbial populations of the kefir grains. In 1963, it was found that the predominant yeast in kefir grains to be *Saccharomyces delbrueckii* (now known as *Torulaspora delbrueckii* [6, 11]), which was associated

---

[2] Martinus Willem Beijerinck was born March 16th, 1851 in Amsterdam [15]. He studied chemistry under Van't Hoff at the Delft Polytechnical School and biology at the University of Leiden. In 1873, he was hired to teach at the Agricultural School in Warffum, but his teaching was found lacking and he lasted only one year [15]. In 1875, he became a part-time teacher at the State Secondary School at Warffum, before taking a teaching position at the Agricultural High School in Wageningen in 1876 [14, 15]. In 1885, he became a microbiologist at the Netherlands Yeast and Alcohol Manufactory in Delft, where he became quite successful [14, 15]. In 1895, he returned to academics when the Dutch government created a special position for him at the Delft Polytechnical School, a position he held until his retirement in 1921 [14, 15]. During his scientific career, he published over 100 articles dealing with a great variety of subjects in the fields of botany, microbiology, and virology. His scientific achievements include fundamental papers on the physiology of luminescent bacteria, the root nodules of *Leguminosae,* and bacterial nitrogen fixation. [14]. He died January 1, 1931 in Gorssel, Netherlands [15].

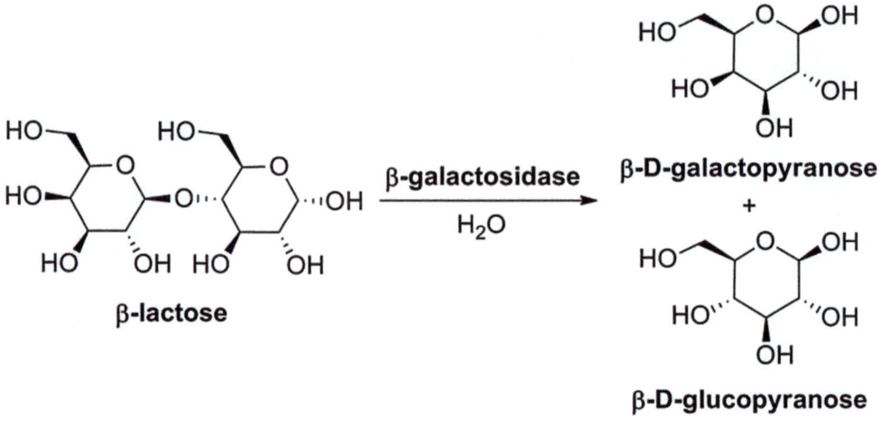

Fig. 5.2 Enzymatic hydrolysis of lactose

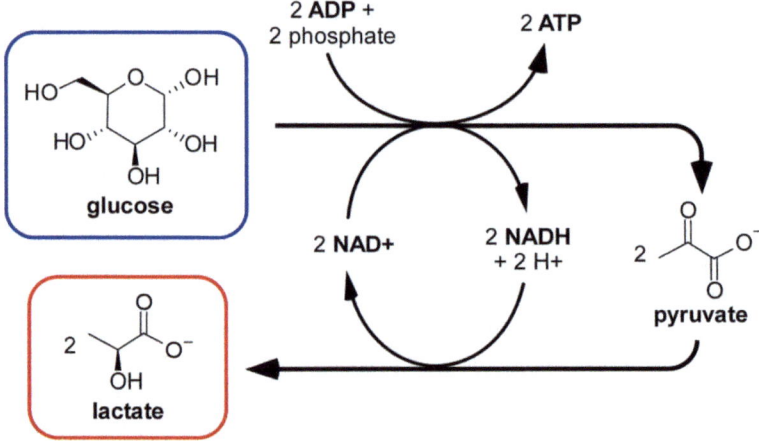

Fig. 5.3 A simplified outline of the lactic acid fermentation process

with *Lactobacillus brevis*. The two organisms depend on each other for survival in milk [11]. Others have reported that *Saccharomyces cerevisiae* has been found to be the most commonly isolated yeast species in kefir grains [9, 10].

A simplified scheme of the lactic acid fermentation[3] of glucose is outlined in Fig. 5.3. As previously discussed for alcoholic fermentation in Chap. 2, each step in Fig. 5.3 actually represents a significant number of individual steps requiring various enzymes. Unlike alcoholic fermentation, however, lactic acid fermentation does not produce CO_2 as a byproduct. Alcoholic fermentation of glucose produced from the hydrolysis of lactose occurs in an identical fashion to that previously

---
[3] Also referred to as lactic fermentation, homolactic fermentation, or glycolysis.

discussed in Chap. 2 and *Saccharomyces cerevisiae* has been reported to facilitate the alcoholic fermentation of galactose by an analogous process [16].

The kefir grains and the exact process of making kefir were often guarded very closely and regarded as a source of wealth. As the kefir grains could be strained, recovered, and reused indefinitely, specific strains could also be passed down from generation to generation.

5.2 Kumis

Kumis[4] is an ancient fermented beverage traditionally made from mare's milk in North Central Asia and Mongolia [6, 12, 17]. Although it is agreed that the horse was domesticated at a later date than other livestock, it is yet unknown when and where horse domestication first took place. Genetic studies of modern domestic horse breeds (*Equus caballus*) imply either multiple domestication events or the breeding of domestic stallions from a single original lineage with captured local juvenile wild mares, but fail to clearly identify the origin of domestication [3]. A prime candidate for the site of domestication of the horse is the Eurasian steppe, specifically the Eneolithic Botai Culture of Kazakhstan. Archaeological investigations of the Botai Culture have revealed the processing of mare's milk in pottery vessels, suggesting the domestication of the animal as early as mid-4th millennium BCE [3]. No specific evidence for the fermentation of mare's milk was found as a part of these investigations, but it is thought that the possibility of such milk fermentation in the Botai Culture is high [3].

The earliest description of the processing of mare's milk was given by the Greek historian Herodotus in 440 BCE. In the following, he describes the processing methods of the Scythians [19]:

> When they had drawn the milk they pour it into wooden vessels hollowed out, and they set the blind slaves in order about the vessels and agitate the milk. Then that which comes to the top they skim off, considering it the more valuable part, whereas they esteem that which settles down to be less good than the other.

Although fermentation is not specifically mentioned, the account of Herodotus is widely believed to be a description of kumis production. Later accounts are more detailed and specifically refer to the production of kumis. One such account was given by 13th century Franciscan monk William of Rubruck[5] who describes the production of the fermented beverage by the Mongols as follows [20]:

---

[4] Also known as *kumiss*, *koumiss*, *kumys*, or *kumyss* [6, 12, 18].

[5] William of Rubruck (ca. 1220–1293) was a Flemish Franciscan monk, missionary, and explorer who set out for the land of the Tartars provided with credentials of King Louis IX of France directed to the Mongol chiefs Sartach and Batu. William returned to Cyprus in 1255 and ultimately presented to the King of France a very clear and precise report of his travels, considered to be one of the great masterpieces of medieval geographical literature, comparable to that of Marco Polo, and quoted by later authors such as Roger Bacon [21].

This cosmos,[6] which is mare's milk, is made in this wise. They stretch a long rope on the ground fixed to two stakes stuck in the ground, and to this rope they tie toward the third hour the colts of the mares they want to milk. Then the mothers stand near their foal, and allow themselves to be quietly milked; and if one be too wild, then a man takes the colt and brings it to her, allowing it to suck a little; then he takes it away and the milker takes its place. When they have got together a great quantity of milk, which is as sweet as cow's as long as it is fresh, they pour it into a big skin or bottle, and they set to churning it with a stick prepared for that purpose, and which is as big as a man's head at its lower extremity and hollowed out; and when they have beaten it sharply it begins to boil up like new wine and to sour or ferment, and they continue to churn it until they have extracted the butter. Then they taste it, and when it is mildly pungent, they drink it. It is pungent on the tongue like râpé wine [i.e., a wine of inferior quality] when drunk, and when a man has finished drinking, it leaves a taste of milk of almonds on the tongue, and it makes the inner man most joyful and also intoxicates weak heads, and greatly provokes urine.

William goes on to also describe kara kumis, or black kumis, the more highly regarded beverage made from the milk of Imperial mares [18, 20].

Kumis is similar to kefir, but is produced from a liquid starter culture, rather than via kefir grains. These starter cultures involve various thermophilic lactobacilli (*Lactobacillus delbrueckii* subsp. *bulgaricus*, *Lactobacillus acidophilus*) and yeasts (*Saccharomyces* sp.) [6, 12]. On average, mare's milk contains 6.4 % lactose by weight, a sugar content that is roughly 30 % higher than that of cow's milk [13]. As a result of this higher sugar content, kumis generally has higher alcohol content than kefir. The alcohol content of kumis varies from 0.6 to 2.5 %, with a lactic acid content of 0.55–1.05 %. As with kefir, kumis is carbonated, and it has been reported to be well saturated with CO_2 [6].

A number of fermented beverages that represent modified forms of kumis have also been produced by various cultures, most of which are made with other animal milks, rather than the traditional mare's milk. Leben (or laban) is one such related drink produced in Lebanon, Iraq, and Egypt [6, 12]. It is made from the milk of buffalo, goat, or cow [12] and is typically diluted down with water to give a weaker beverage [6]. Additional variants of kumis made with cow's milk include *bland* (or blaand) in Shetland and Scotland and *syre* in Iceland [17].

References

1. Bronowski J (1973) The Ascent of Man. Little, Brown and Company, Boston, p 61.
2. Chessa B, Pereira F, Arnaud F, Amorim A, Goyache F, Mainland I, Kao RR, Pemberton JM, Beraldi D, Stear M, Alberti A, Pittau M, Iannuzzi L, Banabazi MH, Kazwala R, Zhang YP, Arranz JJ, Ali1 BA, Wang Z, Uzun M, Dione M, Olsaker I, Holm LE, Saarma U, Ahmad S, Marzanov N, Eythorsdottir E, Holland MJ, Ajmone-Marsan P, Bruford MW, Kantanen J,

---

[6] It is generally agreed that the term *cosmos* used here refers to kumis. It is believed that this is a combination of an incorrect rendering of the intended kumis, combined with a fanciful spelling introduced for the sake of repetition of sound between the two syllables [18].

References

Spencer TE, Palmarini M (2009) Revealing the History of Sheep Domestication using Retrovirus Integrations. Science 324:532–536.
3. Outram AK, Stear NA, Bendrey R, Olsen S, Kasparov A, Zaibert V, Thorpe N, Evershed RP (2009) The Earliest Horse Harnessing and Milking. Science 323:1332–1335.
4. Slavomil V (1994) The archaeology of thirst. J Euro Archaeol 2:299–326.
5. Hornsey IS (2003) A History of Beer and Brewing. The Royal Society of Chemistry, Cambridge, p 8.
6. Roginski H (1988) Fermented Milk. Aust J Dairy Technol 43:37–46.
7. Tamime AY (2002) Fermented milks: a historical food with modern applications—a review. Eur J Clin Nutr 56(Suppl 4):S2–S15.
8. Agata L, Jan P (2012) Production of fermented goat beverage using a mixed starter culture of lactic acid bacteria and yeasts. Eng Life Sci, 12:486–493.
9. Hart E (1883) Kefir, A New Milk-Ferment. Brit Med J 1(1167):925.
10. Motaghi M, Mazaheri M, Moazami N, Farkhondeh A, Fooladi MH, Goltapeh EM (1997) Short Communication: Kefir production in Iran. World J Microbiol Biotechnol 13:579–581.
11. Barnett JA, Lichtenthaler FW (2001) A history of research on yeasts 3: Emil Fischer, Eduard Buchner and their contemporaries, 1880-1900. Yeast 18:363–388.
12. Hesseltine CW (1965) A Millennium of Fungi, Food, and Fermentation. Mycologia 57:149–197.
13. Malacarne M, Martuzzi F, Summer A, Mariani P (2002) Protein and fat composition of mare's milk: some nutritional remarks with reference to human and cow's milk. Int Dairy J 12:869–877.
14. Rouwenhorst RJ, Pronk JT, van Dijken JP (1989) The discovery of β-galactosidase. Trends Biochem Sci 14:416–418.
15. Chung KT, Ferris DH (1996) Martinus Willem Beijerinck (1851-1931), Pioneer of general microbiology. ASM News 62:539–543.
16. Sohngen NL, Coolhaas C (1924) The Fermentation of Galactose by Saccharomyces Cerevisiae. J Bacteriol 9:131–141.
17. Victor Jagielski (1872) Koumiss In The Treatment Of Phthisis. Brit Med J 1(579):124–125.
18. Clark LV (1973) The Turkic and Mongol Words in William of Rubruck's Journey (1253–1255). Journal of the American Oriental Society 93:181–189.
19. Herodotus (1890) The History of Herodotus. Macaulay GC (trans) Macmillan, London, Book IV, verse 2.
20. William of Rubruck (1900) The journey of William of Rubruck to the eastern parts of the world. Rockhill WW (trans) Hakluyt Society, London.
21. Charpentier J (1935) William of Rubruck and Roger Bacon. Geografiska Annaler 17(Suppl):255–267.

Chapter 6
Distillation and the Isolation of Alcohol

As discussed in the previous chapters, the production of alcoholic beverages via fermentation is thought to date back to sometime before 6000 BCE [1–4], yet the separation, isolation, and application of alcohol as a distinct chemical species did not occur until the 12th century CE [4–12]. This of course begs the question: *Why did it take so long*? The principle factor in the long delay between these events was the fact that the most common method for this separation, distillation, was not really developed until the 1st century CE [13–20]. Even so, it still took essentially another thousand years for the successful isolation of ethanol. To understand this additional delay, we first need to briefly introduce the history of distillation methods before returning to its use in the isolation of alcohol.

6.1 A Brief History of Distillation Methods

Distillation is an ancient art and the apparatus used for this process, commonly referred to as a still, is thought to be the earliest specifically chemical instrument [13–15, 20]. Distilling apparatus was introduced by the members of a school of first and second century CE alchemists referred to as the school of Maria the Jewess.[1] Within this group of authors, it is Maria the Jewess who first described this distilling apparatus in the late first century CE, with its development already fairly advanced in her writings. As such, she is generally given credit for its invention [13–20]. Unfortunately, little is known about Maria other than writings ascribed to her, which survive only in quotations by the later alchemist Zosimos [13, 14, 18]. One such quotation describes a three-armed still, known as a *tribikos*, which has been reported as follows [14, 18]:

---

[1] Also sometimes given as Mary or Miriam, Maria the Jewess was an alchemist of the first century CE who wrote on practical chemical aspects and described various types of early laboratory apparatus [8, 13, 14, 16]. In addition to the still, she is commonly credited with the invention of the water-bath and the *kerotakis* apparatus [8, 13, 16, 17, 19]. For this reason, the water bath was referred in Latin to as the '*balneum Mariae*' and later in French as the '*bain Marie*' [8, 16, 17, 19]. Maria is alleged by some to be Miriam, the sister of Moses [17, 19, 20].

> I shall describe to you the tribikos. For so is named the apparatus constructed from copper and described by Mary, the transmitter of the art. For she says as follows:
> Make three tubes of ductile copper[2] a little thicker than that of a pastry-cook's copper frying-pan: their length should be about a cubit and a half. Make three such tubes and also make a wide tube of a handsbreadth width and an opening proportioned to that of the still-head. The three tubes should have their openings adapted like a nail to the neck of a light receiver, so that they have the thumb-tube and the two finger-tubes joined laterally on either hand. Towards the bottom of the still-head are three holes adjusted to the tubes, and when these are fitted they are soldered in place, the one above receiving the vapour in a different fashion. Then setting the still-head upon the earthenware pan containing the sulphur, and luting the joints with flour paste, place at the ends of the tubes glass flasks, large and strong so that they may not break with the heat coming from the water in the middle.

Representative illustrations of distillation apparatus from Hellenistic manuscripts of this time period are given in Fig. 6.1. Various authors remind us that these figures come from copies dating centuries after that of the originals. As such, we do not know how many times these figures were recopied, nor what errors, distortions, or modifications may have been introduced in the process [13, 14]. However, they do seem to correspond to the descriptions given in the texts. Of course, the simplest form of the still utilizes only a single receiver, of which a later woodcut is shown in Fig. 6.2.

The early traditional still consisted of three components (Fig. 6.3): the distillation vessel (*cucurbit*[3]) [6, 15–17, 21], the still-head (*ambix*) with an attached delivery tube (*solen*), and the receiving vessel (*bikos*) [15–17, 21]. Thus, the *tribikos* described by Maria the Jewess refers to the fact that this still utilized three receiving vessels. For such *tribikos* and *dibikos* apparatus, it is unclear whether these were meant to be used for two or more fractions or the simultaneous collection of distillate in multiple receivers. In terms of practical technical details, simultaneous collection would be the simplest possible use. These *bikos* receiving vessels were generally comprised of a small body with a long thin neck and such flasks in general were also called a *phial* [16, 21].

As demonstrated by the quotation of Maria the Jewess above, the early stills were made from a mixture of primarily earthenware (with the interior glazed), copper, and glass. Initially, however, glass was generally limited to just the receiving vessels [12, 14, 15, 17, 21]. As glass industries evolved, it became more common to use glass for first the *ambix* and then later for both the *cucurbit* and *ambix* [12, 15, 21]. One of the difficulties encountered with the use of glass in still components, however, was the breaking of the vessels during heating because the glass was typically irregular and of poor quality. This issue is reflected in following portion of the quotation of Maria the Jewess by Zosimos [14, 18]:

> …glass flasks, large and strong so that they may not break with the heat coming from the water…

---

[2] Sometimes given as bronze rather than copper [13, 17, 19].
[3] From the Latin *cucurbita* meaning "gourd" [19]. The *cucurbit* is sometimes called a *bikos* as well. Alternate terms also include *lopas* and *botarion* [16].

6.1 A Brief History of Distillation Methods

Fig. 6.1 Depictions of early chemical apparatus from Hellenistic manuscripts [14]

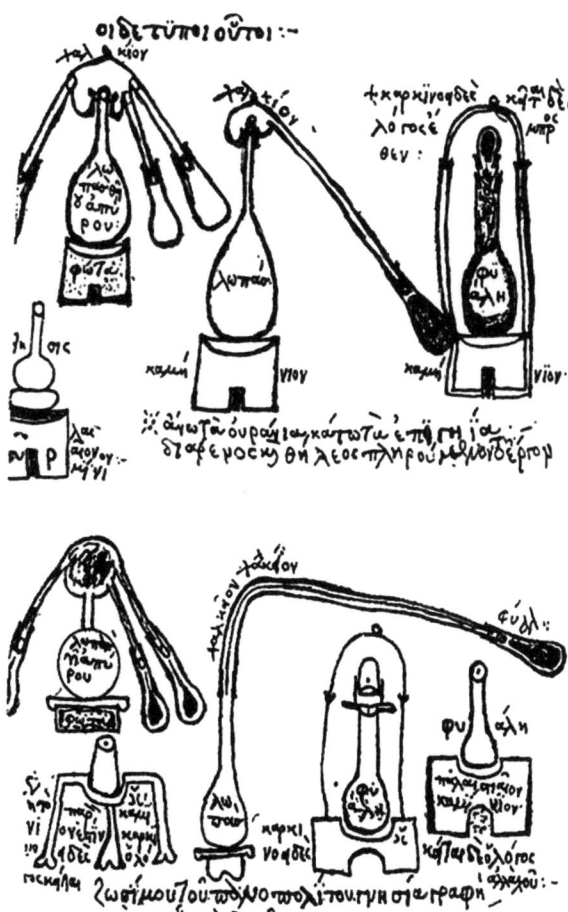

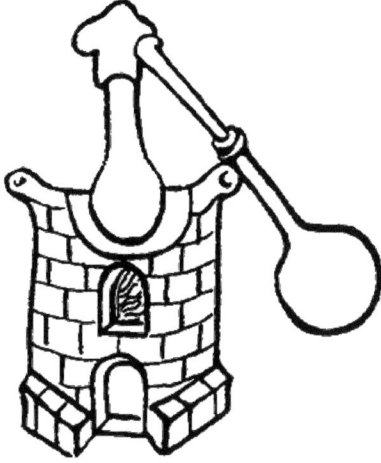

Fig. 6.2 Medieval woodcut of a basic still

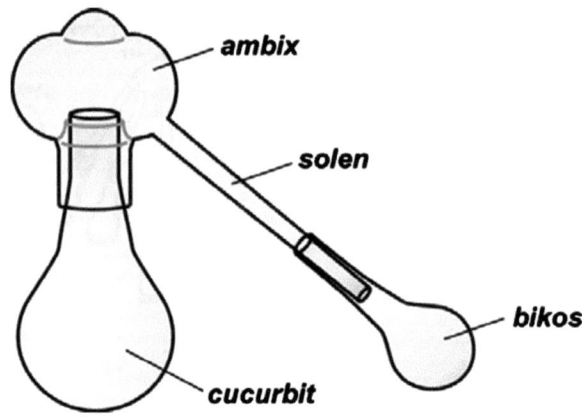

Fig. 6.3 Basic components of the early still

Pliny the Elder also warned of similar issues, stating that [22]:

Glass is unable to stand heat unless a cold liquid is first poured in.

Early glasses up through the Roman period were primarily soda-lime glasses that utilized a simple soda-lime-silica composition that varied depending upon the availability of raw materials [21, 23]. Within this composition, the calcium of the lime acted as a stabilizer for the glass to counteract the high solubility of the sodium contained within the glass [24]. Unfortunately, the importance of lime was not initially recognized and it was not intentionally added as a major constituent before the end of the 17th century [25, 26]. Prior to that time, all calcium content in early glasses was a result of impurities in either the silica or soda sources [21, 23]. As a consequence of the high soda content and low lime content, such early glasses generally suffered from both a lack of durability under rapid temperature changes and poor chemical resistance [15, 21, 27, 28]. For simple soda glasses, the thermal expansion of the glass increases with soda content [29]. Therefore, the high thermal expansion from high soda content coupled with the physical defects common in early glass resulted in glasses with low thermal durability and a tendency to break under rapid heating [21, 23]. It has been recognized that the distillation efforts prior to the 12th century were quite limited by both the poor quality of glass and ineffective cooling methods to effectively collect the condensing material [10].

In an effort to overcome the poor heat stability of the glass, vessels with thick walls were typically used. In addition, a thick coating (up to two or three fingers) of clay was often applied to the exterior of the cucurbit to further strengthen the glass under heating [12, 15, 16, 21, 30]. The coating of clay helped reduce breaking, but the poor heat transmission of such a coating resulted in unnecessarily long preheating periods. The combination of poor heat transmission and the lack of efficient cooling thus made it difficult to distill volatile liquids such as alcohol [15, 21, 31].

The various pieces of the still were fixed together using a *lute* to seal the joints between the multiple components, particularly between the curcurbit and the still-

6.1 A Brief History of Distillation Methods

head. The lute needed to be a material of adequate plasticity and a variety of materials have been reported to be used for this purpose. Some authors have stated that an early lute was a flour paste [12, 17]. Other reported materials have included fat, wax, clay, lime mixed with egg white, and gypsum or clay mixed with oil [9, 15, 16, 31]. In later times, a special luting known as *philosophers' clay, clay of wisdom, lute of the philosophers*, or *lutum sapientiae* was highly recommended [12, 16, 17]. According to one such recipe, this lute consisted of two-thirds stone-free clay and one-third dried dung mixed with chopped hair [17]. Other such recipes included either a mixture of lime ($CaCO_3$), caseine, glue, and white of egg [16], or a mixture of common clay, potter's clay, horse manure and chopped straw (or glass powder), quicklime (CaO), powdered bricks, white lead ($Pb_2(CO_3)(OH)_2$) and egg whites [12].

By the 10th century, the term *ambix* came to be applied not just to the still-head, but to the still as a whole [16, 17, 19]. As distillation apparatus were transmitted to the Islamic philosophers during the 7th–8th centuries CE, the term *ambix* was transformed through the addition of the Arabic article *al-* to become *al-anbîq*, which eventually became *alembicus* and *alembic*[4] [14–17, 19, 21]. As with the use of ambix by this point, the term *alembic* was commonly used to refer to both the still-head and the still as a whole [12, 14, 15, 21]. Because of the Arabic origin of the word *alembic*, some authors mistakenly attribute the discovery of distillation methods to the Islamic philosophers [4].

Initial efforts to improve cooling methods were to cool the delivery tube (*solen*) with wet sponges or rags. As the delivery tube was now typically cooler than the still-head, condensation occurred primarily in the delivery tube. Because of this, the typical medieval alembic no longer contained an inner rim for collecting the condensate within the still-head (Fig. 6.4) and the art of fabricating the old traditional form was gradually lost [12, 15, 21]. One of the earliest references to distilled alcohol is found in the writings of Magister Salernus[5] [6, 9, 12]. As such, it is believed by some that he may have pioneered the cooling of the solen to effect condensation outside the still-head [21, 31].

Beginning in the 13th century, the prospering Venetian glass industry began blending Roman and Syrian glassmaking methods which resulted in a significantly improved glass by the mid to later 13th century [21, 32–34]. The marked improvement in glass technology was largely due to careful selection of reagents, as well as the introduction of new processes for the purification of these reagents prior to their use in glass production [35]. Perhaps the most critical factor, however, was due to the choice of alkali used to make the glass. Up through the Roman period, the most common source of soda was *natron*, a term used to describe a mixture of evaporite minerals comprised primarily of sodium sesquicarbonate,

---

[4] Alternate forms included *alembik, alembyk, alembike, alembyke, alimbeck, alembeke, alimbecke, alimbeck*, and *limbick* [12].

[5] Usually referred to with the title of Magister or Master, Salernus (d. 1167) was a physician of the School of Salerno and lived at Salerno between 1130 and 1160. His writings included a summary of pathology and therapeutics [8, 9, 11, 17].

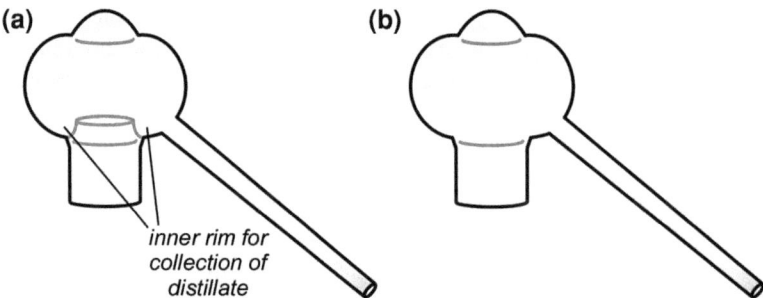

Fig. 6.4 Early still-head utilizing an inner rim for collection of distillate (**a**) and a later still-head without an inner rim (**b**)

$Na_2CO_3 \cdot NaHCO_3 \cdot 2H_2O$. The Venetians, however, utilized plant ashes imported from the Levant, which in addition to Na_2CO_3, contained significant amounts of magnesium and calcium [35]. The resulting change in glass composition resulted in a new glass that exhibited both higher chemical durability and reduced thermal expansion, while the new purification methods removed insoluble, non-fusible components from the resulting glass products which would have acted as points of stress during heating [27–29, 35, 36].

With the introduction of the improved Venetian glass, it became more and more common to use both glass cucurbits and alembics [12, 15, 21]. The common fabrication of such glass components then allowed the investigation of more versatile approaches to cooling. These ideas culminated in a design that passed an elongated solen through a tub of water for more efficient cooling of the delivery tube as shown in Fig. 6.5 [9, 11, 12, 15, 21]. This idea was introduced during the late 13th century in the writings of Taddeo Alderotti[6] of Florence (ca. 1210–1295), who is commonly thought to be its inventor [9, 10, 12, 17, 37]. In his *De virtutibus aquae vitae*, Alderotti describes the distillation of wine using a tightly luted alembic and receiver. The alembic is specified to utilize an elongated solen consisting of a *canalem serpentinum*, along with a cooling trough and regular supply of fresh cooling water [9].

---

[6] Taddeo Alderotti (also known as Taddeo degli Alderotti, Taddeo de Firenze, and Thaddeus Florentinus [9, 37–39]) was born in Florence [17, 38], although the date of his birth is unknown [41]. His earliest biography states that he died in his 80s and thus it is thought that he was born between 1206 and 1215 [38]. It is said that he was brought up in poverty and thus his education could not begin until relatively late [40]. It has been proposed that his early studies were fostered by the Franciscans sometime between the mid-1230s and the early 1260s [38], after which he started teaching medicine and logic at the University of Bologna as early as 1260–1264 [38, 40, 41], where he was among the very first teachers of the newly founded university. He was an author on anatomy and medicine [9, 38, 40, 41] and advocated the close association of medicine with Aristotelian natural philosophy and defended the validity of medicine as a science [41]. He discusses the distilling of alcohol and its medicinal value in his *De virtutibus aquae vitae* [9], the final section of his *Consilia medicinalia* written about 1280 [38, 39]. He died in Bologna in 1295 [17, 38, 41], although dates of 1292 and 1303 have been reported by some authors [9, 38, 40, 42].

Fig. 6.5 External cooling trough as depicted in the treatise of Johann Wenod, 1417–1418 [37]

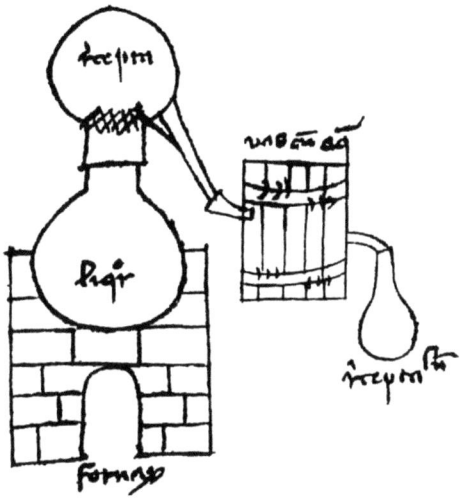

The earliest known pictorial representation of this new cooling method (Fig. 6.5), however, was not given until about 1420 by Johann Wenod,[7] a physician in Prague [12, 15, 37, 43, 44]. Unfortunately, the form and nature of the cooling tube within the tub is not detailed in Wenod's illustration, and he only gives the notation "*vas cum aqua*" (vessel with water) above the cooling tub [37]. Neither are details more forthcoming in Wenod's writings, where he seems to believe that he is discussing something already in general use. As such, he does not explicitly state any coiling or lack thereof of the tube within the cooling tub and only stresses the length of the solen [37].

Based on the description *canalem serpentinum* ("serpentine channel") given by Alderotti, however, it was thought that his cooling apparatus wound 'worm-like' through the cooling trough as shown by many later pictures (Fig. 6.6) [8, 12]. As such, this type of cooling apparatus is referred to as a "wormcooler" and would be the precursor to the modern cooling coil commonly used in current reflux and distillation apparatus. It is thought that through the use of his "*canale serpentinum*" run through a cooling trough with a regular supply of fresh cooling water, it

---

[7] Other than the fact that he was a physician in Prague around the period of 1420, and is believed to have originally come from Altenburg in Germany, very little is known about Johann Wenod [37, 44]. The only known documentation is a brief medical treatise on the treatment of urinary stone disease and related drug preparations, which was included in a manuscript written in 1417–1418 [44]. This treatise was discovered by Karl Sudhoff in the Leipzig University Library (Ms. 1775), the details of which he published in 1914 [37, 44]. From Wenod's choice of wording in his writing, Sudhoff felt Wenod to be German, but states uncertainty about the correct surname of the author. While he favored either 'Venod' or 'Wenod', he states that these could also be a Latinized form of 'Wende' [43]. Although Sudhoff gives his first name as Johann [37, 44], which would be consistent with a German heritage, references by most authors use the English equivalent 'John' [12, 15, 43].

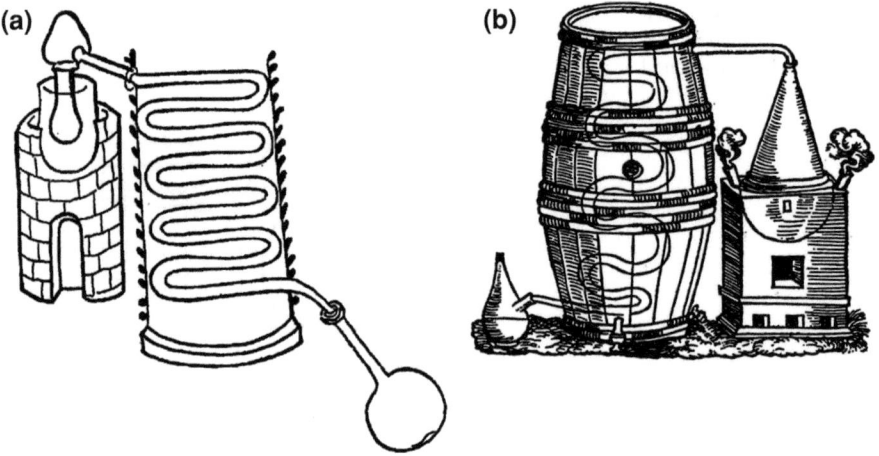

Fig. 6.6 Illustrations of the worm-like nature of the "wormcooler" cooling coil from: **a** Philipp Ulstadt's *Coelum Philosophorum seu De Secretis Naturae Liber*, 1525 and **b** Walter Ryff's *Neu Gross Destillierbuch*, 1556

should have been possible for Alderotti to obtain 90 % alcohol by fractional distillation [11, 12, 17].

The impact of the improved Venetian glass and the growing glass industry on the evolution of distillation apparatus led not only to the development of new, improved glass-based components such as the wormcooler, but also to the eventual move away from the use of even simple earthenware components to new stills fabricated completely from glass [15]. As all-glass distillation apparatus become more common, the distillation flask and alembic were eventually combined and blown or cast in one piece. This new form was called the *retort* (from Latin *retortus*, "bent back", Fig. 6.7) and was introduced in the early 14th century [12]. The retort was especially well-suited for distillation at higher temperatures when the lutes in a typical alembic would begin to fail [43]. Distillation via a retort was often referred to as *destillatio ad latus* ("side-wards distillation") [12].

Two later still designs, the *Rosenhut* and the *Moor's head*, both focused on the still-head instead of the solen in efforts to increase effective cooling during distillation (Fig. 6.8) [12, 45, 46]. The Rosenhut (German) or Rozenhoed (Dutch, both literally meaning "rose hat") is thought to be the earlier design of the two as it was illustrated in its fully developed form in 1478 [12], but essentially disappeared by the end of the 16th century [45]. The Rosenhut consisted of a high conical, air-cooled alembic and was very common in early apparatus used for making liqueurs [12, 46]. This modified alembic was typically fitted to a wide-mouthed cucurbit and is thought to have been built with an inner rim to collect the distillate, although this has never been shown in illustrations of the Rosenhut [12]. While it was most common to apply glass distillation components by this point, the Rosenhut was frequently made from metals such as lead and copper as the high thermal conductivity of the metals resulted in superior air cooling [12, 46].

6.1 A Brief History of Distillation Methods

Fig. 6.7 Illustration of a basic early retort (**a**) and a photo of a modern glass retort (**b**)

Fig. 6.8 Illustration of the Rosenhut from Michael Puff von Schrick's *Hienach volget ein nüczliche materi von manigerley ausgepranten wasser*, 1478 (**a**) and the Moor's head from Hieronymus Brunschwyck's *Liber de arte distillandi de Compositis*, 1512 (**b**)

In contrast to the air-cooled Rosenhut, the *Mohrenkopf* or "*Moor's head*" enclosed the still-head in a basin or container which was filled with cooling water (Fig. 6.8b). The Moor's head is thought to be an invention of the later 15th century and was typically made of glass (although pottery is also said to have been used) [12, 46]. From its name, it is believed by some that the development of the Moor's head may have involved influence from the world of Islam as the term "Moor" usually referred to medieval Muslims that inhabited primarily Morocco and surrounding African regions (western Algeria, Western Sahara, Mauritania), as well as the Iberian Peninsula, Sicily and Malta. However, the term has also been used in a broader sense to refer to Muslims in general and it has been suggested that its use here could refer to someone from the "further Indies" and may have been influenced by the Chinese still, which also utilized a water-cooled head [11]. In comparison to the wormcooler, it has been stated that the Moor's head gave distillates of lower quality [47].

By this point, glass had become by far the preferred medium for chemical glassware, particularly that of equipment for distillation. More importantly, 16th-century authors such as Hieronymus Brunschwig (1450–1512)[8] and Conrad Gesner (1516–1565)[9] specified not only glass distillation components, but preferably those constructed of Venetian glass [44, 48]. Brunschwig even stated that the distillation vessels [48]:

> ...must be made of venys [Venetian] glasse bycause they shoulde the better withstande the hete of the fyre.

6.2 Distillation of Wine

The availability of more advanced still designs that provided more effective cooling then made it possible to prepare alcohol distillates of high concentration [46]. Available evidence has led to the current belief that the initial isolation of

---

[8] Hieronymus Brunschwig (also given as Brunschwyck, Braunschweig, or Brunschwijg) was a German surgeon and a native of Strassburg [45, 51, 52]. He is believed to have been born ca. 1450 [45, 52], although others have given earlier dates of 1430 [51] or 1440 [52]. He was descended from the Sauler (or Saler [51]) family of Strassburg and studied medicine at Bologna, Padua and Paris [45]. He is most well-known for his works on the art of distillation, most importantly *Liber de arte distillandi de simplicibus* (1500, commonly known as the *Small Book of Distillation*) and *Liber de arte distillandi de Compositis* (1512, commonly known as the *Large Book of Distillation*) [45, 51, 52]. He died at the end of 1512 or the beginning of 1513, at the age of 60 [45, 51, 52]. Others, however, have given his death as 1533 or 1534 [51, 52].

[9] Conrad Gesner, also referred to as Evonymus Philiater, was born in Zurich on March 26, 1516 [51, 53]. He was destitute at an early age due to his father's death, which forced him to take an assistant's position in Strassburg. In 1535, he took a teacher's post at Zurich and studied medicine, before becoming a professor of Greek philology at Geneva in 1537 [53]. He finally finished his medical study at Basel in 1541 [53] and then returned to Zurich as a private physician and professor of physics and natural history [51]. He was a voluminous writer and died on December 13, 1565 [51, 53].

6.2 Distillation of Wine

alcohol occurred in southern Italy during the 12th century,[10] most likely at the School of Salerno, an important medical school [4–13, 49]. Of course, it is possible that the isolation of alcohol could have occurred at an earlier date, and in fact some authors do argue for its earlier discovery by the Muslim philosophers [50]. While it is believed that the Arab alchemists studied the distillation of wine prior to the 12th century [11, 50], no convincing evidence of the isolation of ethanol prior to that of the western sources has yet been presented [12, 54]. It has been pointed out that the common view is that the Arabs did not find the products obtained from the distillation of wine very interesting [11]. Such a lack of interest would make sense if the distillates still contained high water content as would be consistent with the ineffective isolation of the volatile alcohol products.

This assignment to 12th century Italy is strengthened by the fact that one of the earliest direct recipes for isolation of alcohol is contained in the writings of Magister Salernus [7, 9, 10, 12]. Earlier treatises from Salerno from the time period of 1100–1150 CE discuss the preparation of "beneficial waters" by distillation [6, 11, 12], but his writings were the first from the School of Salerno to directly mention alcohol [12]. This recipe for *aqua ardens* (burning water) was said to be made after the fashion of rosewater as follows [7]:

> Place in the cucurbita one pound (white, or) red wine, one pound powdered salt, four ounces native sulphur, four ounces of tartar (from wine). The liquid distilling is collected. A cloth saturated with this liquid will maintain a flame without suffering injury. Cotton does the same without loss of substance.

Rosewater was one of the primary distillation products of the Islamic philosophers, which was produced from rose petals as early as the 9th century. The flowers were macerated with water and the essential oils distilled over with the water using a conventional alembic [55].

Another recipe from the same time period is found in the *Mappae Clavicula*, which may predate the recipe by Salernus [7, 10, 11, 49]. This Medieval Latin text, the title of which translates as "Little Key to Painting", consists of a compilation of recipes for a number of crafts including metalwork, dyeing and other chemical arts [7, 11]. It is thought that the earliest version of this text dates to ~ 820 CE, but available manuscripts consist of only 10th and 12th century versions [11, 49]. The alcohol recipe is found in the 12th century version, but is not contained in the earlier version [10, 11, 49]. The recipe from the *Mappae Clavicula* was written with portions given in the form of a cryptogram, as shown in the original Latin [7]:

> Ad bonum argentum solidandum nedium oboli. De commixtione puri et fortissimi xknk cum iij qbsuf tbmkt, cocta in ejus negocii vasis fit aqua quae accensa flammans incombustam servat materiam.

---

[10] The exact date is unknown and various dates have been reported, including ~ 1100 CE [4, 10, 12], ~ 1130 CE [7], and ~ 1150 CE [6]. The timespan of 1150–1170 CE, however, has the strongest support from available evidence [6, 11].

The solution to this cryptogram was originally solved by Berthelot [7, 11] in which each letter of the transformed words could be revealed by substituting the given letters by the letter directly preceding it in the alphabet [7]. The solved and translated text then gave the following [7, 10]:

> By mixing pure and strongest wine with three parts of salt and heating in a vessel customary for that purpose, a water is produced which when kindled inflames, [yet] leaves the material unburned.

An additional recipe has been reported to have been found in a 12th century parchment recovered from Weissenau, a south German monastery. The translated recipe was given as [8]:

> Place in the cucurbita one pint wine and one pound red sulphur salt, or also four ounces living sulfur, which has been roasted in an ordinary pot and four ounces tartar, which you add all together and then close with a cover, the watery part which flows down through the nose of the cover you collect.

After the 13th century, recipes for preparing alcohol are frequent in the available literature [10].

The reason for the late discovery of alcohol was partly due to the long preheating period coupled with inefficient cooling during distillation. Some emphasize the dependence of alcohol's eventual isolation on the improvements in cooling methods [11] and such improvements in still design perhaps had the greatest overall impact. Of course, none of the early recipes above give details on the distillation methods used, particularly the nature of any cooling methods, and thus it is unknown if these initial successes in the isolation of alcohol utilized methods for cooling the solen. It has been pointed out that it is conceivable to distill alcohol in the ancient cucurbit and alembic without cooling the solen, or even just cooling the alembic, but only if the temperature could be regulated carefully [12]. However, the common use of earthenware or clay-coated glass curcurbits resulted in poor heat transmission through the curcurbit material, usually resulting in long digestion periods before distillation and excessive temperatures that drove off the low boiling fractions, thus making it difficult to distill volatile liquids such as alcohol [12, 15, 21, 31]. Thus, the successful distillation of alcohol would realistically require at a minimum either more effective cooling of the solen or curcurbits constructed of the improved Venetian glass, with the best case being a combination of the two factors. Without such critical factors, most refined alcoholic distillates separated by the early stills contained so much water that they would not burn, thus making it difficult to differentiate these distillates from normal water [11, 12].

Of course, the secret of the success in the 12th century was not just better cooling and fractional distillation, but also the addition of a variety of salt substances (NaCl, potassium tartrate, K_2CO_3, etc.) as illustrated in the various recipes above. The added salts acted by absorbing a portion of the wine's water content, thus increasing the concentration of alcohol and making it easier to isolate via distillation [10, 12]. It is thought that this practice was perhaps influenced by the suggestions of Islamic philosophers that something to absorb one nature should be added in attempting to purify another nature [10].

These early solutions distilled from wine-salt mixtures were referred to as either *aqua ardens* (burning water) [6, 10, 11, 56, 57] or *aqua flamens* (flaming water) and generally had such low alcohol content that they burned without producing noticeable heat [7, 8, 11, 12, 56]. The combination of pretreatment with salts along with more efficient cooling methods ultimately produced alcoholic distillates containing less than 35 % water and 'absolute' alcohol could be obtained via fractional distillation [12].

Monks had long produced wines for sacramental purposes, and were thus many of the first distillers in the west. Initially, it was thought by some that alcohol was the *quintessence*, the fifth element of Aristotle that made up the heavens. As such, they studied its properties intensively. Alcohol looked like water, but burned with a blue, gemlike flame, a perplexing contradiction as all knew that the nature of water was to extinguish fire, not to burn. In addition, when consumed, it produced a quite pleasing intoxication. An alchemical text ascribed to Raymond Lull[11] described alcohol as follows [58]:

The taste of it exceedeth all other tastes, and the smell of it all other smells

The separation of alcohol from wine was viewed to be analogous to the separation of the soul from an impure body and for this reason alcohol was thought to be the "spirit" of the wine and the remaining residue was called the *caput mortum* (dead body). Very strong alcohol distillates were named *aqua vitae* (water of life) [10, 11] by authors such as Arnald of Villanova[12] [8, 12, 59] and the name still

---

[11] Also given as Ramón Lull or Raymond Lully [60]. Considerable obscurity surrounds the life of Raymond Lull, whose name became famous in the 13th century [56, 60]. Many of the works attributed to him are now believed have been written by others. Lull is said to have been descended from a noble Spanish family and was born in Majorca about 1235 [60] (ca. 1232–1236 [56, 61]). He first devoted himself to the study of science, but was converted in 1263 [61] at which point he became a monk and eventually a missionary [56, 60, 61]. According to tradition, he took up the study of alchemy from the desire to cure a girl who was suffering from cancer. Afterward he wandered through Europe to acquire further knowledge of the art. He is said to have made alcohol by distillation and to have known how to dehydrate it by the aid of potassium carbonate. In his later years, he again devoted himself to missionary work and left to preach in Africa. It is there in Algeria that he is said to have died in 1315 (at age 77 [60]), a martyr to his faith [56, 61].

[12] Arnald of Villanova (also given as Arnau de Vilanova [62], Arnaldus de Vill Nova [60, 61], or Villanovanus [61]) was born near Valencia about 1240 (c. 1234–1250) [8, 9, 61]. Others have given his birthplace as Villeneuve-Loubet, near Avignon [60, 62, 63]. He studied medicine at Naples [8, 9, 60] and Paris, where he took the degree of Master of Arts, and completed his studies at Montpellier [60, 63]. He was a Catalan physician and self-styled prophet [61, 62, 64], and was a professor at Montpellier at least up to 1309 [8, 9]. He translated medical works from Arabic into Latin and had some knowledge of Greek and Hebrew. He realized the value of natural science and suggested that it should be given more importance in education [61]. He was a famous medical practitioner, who was consulted by kings and popes [8, 9, 60–62], and is considered by some to be one of the most extraordinary personalities of medieval times. He also had difficulties with the French Inquisition, first in 1299 and again in 1304 [9, 60, 63]. He had a large number of

survives in the modern words aquavit (Scandinavian), eau-de-vie (French), whiskey (Scottish), and vodka (Slavic). Arnald of Villanova said the following on the choice of the name *aqua vitae*:

> This name is remarkably suitable, since it is really a water of immortality. It prolongs life, clears away ill-humours, revives the heart, and maintains youth.

As discussed in Chap. 1, our modern term alcohol was not used to refer to these distillation products until the 16th century, when its use can be found in the writings of Paracelsus [7, 57]. The term alcohol then gradually replaced the older terms of *aqua ardens*, *aqua flamens*, and *aqua vitae* [57].

Although the description of *aqua vitae* is sometimes given as absolute alcohol, it is important to remember that the highest concentration of alcohol that can achieved by conventional distillation of aqueous solutions is 95 %. This is due to the fact that the 95 % ethanol: 5 % water mixture represents what is called a minimum-boiling *azeotrope*. Such an azeotrope has a fixed composition, a fixed boiling point, and cannot be separated (or its composition altered) by distillation. In all respects an azeotrope acts like a pure liquid.

By 1288, the study of alcohol had perhaps gone a bit too far, because the Dominican provincial chapter at Rimini declared it forbidden for brethren to possess the instruments by which they make the water called *aqua vitae* [11, 56]. This was at least partially because its production was associated with the apothecaries, who were the first to produce alcohol on a large scale [12], and the secular nature of the apothecary trade was considered inappropriate or beneath such men of the Church [11]. Nevertheless, some monks still continued to produce alcoholic beverages, examples of which include the oldest known liqueur *Benedictine*, invented by Dom Bernardo Vincelli in 1510 [56]. This liqueur is made through the extraction and distillation of a variety of herbs with alcohol. A related liqueur *Chartreuse* was then later made by the Carthusian monks [12].

References

1. Lambert JB (1997) Traces of the Past. Unraveling the Secrets of Archaeology through Chemistry. Addison-Wesley, Reading, MA, pp 134–136.
2. Hornsey IS (2003) A History of Beer and Brewing. The Royal Society of Chemistry, Cambridge, pp 9–20.
3. McGovern PE, Hartung U, Badler VR, Glusker DL, Exner LJ (1997) The beginnings of winemaking and viniculture in the ancient Near East and Egypt. Expedition 39:3–21.
4. Vallee BL (1998) Alcohol in the Western World. Sci Am 279(June):80–85.

(Footnote 12 continued)
writings ascribed to him [61, 64], most of them dealing with medical subjects and he was one of the first Latin writers to insist upon the virtues of alcohol. Other works dealt with chemistry, astrology, magic and theology [61]. He died at sea towards the end of 1311 [9, 61, 62, 64] (or 1313 [60, 63]) while on the way from Naples to Genoa, where he was buried [60, 61, 63].

References

5. von Lippmann EO (1912) Zur Geschichte des Alkohols und seines Namens. Angew Chem 25:2061–2065.
6. von Lippmann EO (1920) Zur Geschichte des Alkohols. Chem Ztg 44:625.
7. Stillman JM (1924) The Story of Early Chemistry. D. Appleton and Co., New York, pp 184–192.
8. Liebmann AJ (1956) History of Distillation. J Chem Educ 33:166–173.
9. Forbes RJ (1970) A Short History of the Art of Distillation. E. J. Brill, Leiden, pp 55–65.
10. Leicester HM (1971) The Historical Background of Chemistry. Dover Publications, Inc., New York, pp 76–77.
11. Gwei-Djen L, Needham J, Needham D (1972) The Coming of Ardent Water. Ambix 19:69–112.
12. Forbes RJ (1970) A Short History of the Art of Distillation. E. J. Brill, Leiden, pp 76–98.
13. Talyor FS (1937) The Origins of Greek Alchemy. Ambix 1:30–47.
14. Taylor FS (1945) The Evolution of the Still. Ann Sci 5:185–202.
15. Rasmussen SC (2012) How Glass changed the World. The History and Chemistry of Glass from Antiquity to the 13th Century. Springer Briefs in Molecular Science: History of Chemistry, Springer, Heidelberg, pp 51–65.
16. Forbes RJ (1970) A Short History of the Art of Distillation. E. J. Brill, Leiden, pp 17–24.
17. Holmyard EJ (1990) Alchemy. Dover Publications, New York, pp 47–54.
18. Talyor FS (1992) The Alchemists. Barnes & Noble, New York, pp 39–46.
19. Holmyard EJ (1956) Alchemical Equipment. In: Singer C (ed) A History of Technology, Vol. 2. Clarendon Press, Oxford.
20. Stillman JM (1924) The Story of Early Chemistry. D. Appleton and Co., New York, p 151.
21. Rasmussen SC (2008) Advances in 13th Century Glass Manufacturing and their Effect on Chemical Progress. Bull Hist Chem 33:28–34.
22. Roger F, Beard A (1948) 5,000 Years of Glass, J. B. Lippincott Co., New York, p 233.
23. Rasmussen SC (2012) How Glass changed the World. The History and Chemistry of Glass from Antiquity to the 13th Century. Springer Briefs in Molecular Science: History of Chemistry, Springer, Heidelberg, p 4.
24. Lambert, JB (2005) The 2004 Edelstein Award Address, The Deep History of Chemistry. Bull Hist Chem 30:1–9.
25. Turner, WES (1956) Studies in Ancient Glasses and Glassmaking Processes. Part III. The Chronology of the Glassmaking Constituents. J Soc Glass Technol 40:39–52T.
26. Turner WES (1956) Studies in Ancient Glasses and Glassmaking Processes. Part V. Raw Materials and Melting Processes. J Soc Glass Technol 40:277–300T.
27. Philips CJ (1941) Glass: The Miracle Maker. Pitman Publishing Corporation, New York, pp 43–44.
28. Dimbleby V, Turner WES (1926) The Relationship between Chemical Composition and the Resistance of Glasses to the Action of Chemical Reagents. Part I. J Soc Glass Technol 10:304–358.
29. English S, Turner WES (1927) Relationship between Chemical Composition and the Thermal Expansion of Glasses. J Am Ceram Soc 10:551–560.
30. Forbes RJ (1970) A Short History of the Art of Distillation. E. J. Brill, Leiden, p 114.
31. Gies F, Gies J (1994) Cathedral, Forge, and Waterwheel. Technology and Invention in the Middle Ages. HarperCollins Publishers, New York, p 163.
32. Cummings K (2002) A History of Glassforming. A & C Black, London, pp 102–133.
33. Sarton G (1947) Introduction to the History of Science, Vol. III, Part I. The William & Wilkins Co., Baltimore, pp 170–173.
34. Jacoby D (1993) Raw Materials for the Glass Industries of Venice and the Terraferma, About 1370 - About 1460. J Glass Studies 35:65–90.
35. Rasmussen SC (2012) How Glass changed the World. The History and Chemistry of Glass from Antiquity to the 13th Century. Springer Briefs in Molecular Science: History of Chemistry, Springer, Heidelberg, pp 44–48.

36. Dimbleby V, Muirhead CMM, Turner WES (1922) The Effect of Magnesia on the Resistance of Glass to Corroding Agents and a Comparison of the Durability of Lime and Magnesia Glasses. J Soc Glass Technol 6:101–107.
37. Sudhoff K (1914) Weiteres zur Geschichte der Destillationstechnik. Arch Gesch Naturw Techn 5:282–288.
38. Siraisi NG (1981) Taddeo Alderotti and His Pupils. Two Generations of Italian Medical Learning. Princeton University Press, Princeton, pp 27–42.
39. von Lippmann EO (1914) Thaddäus Florentinus (Taddeo Alderotti) über den Weingeist. Arch Gesch Med 7:379–389.
40. George S (1931), Introduction to the History of Science, Volume II From Rabbi Ben Ezra to Roger Bacon in Two Parts. The Williams & Wilkins Company, Baltimore, Part II, pp 1086–1087.
41. Siraisi NG (1977) Taddeo Alderotti and Bartolomeo da Varignana on the Nature of medical Learning. Isis 68:27–39.
42. Sarton G (1964) A History of Science. Ancient Science through the Golden Age of Greece. John Wiley & Sons, New York, p 381.
43. Partington, JR (1998) A History of Chemistry, Martino Publishing, Mansfield Centre, CT, Vol. 2, p 266.
44. Sudhoff K (1914) "Vera cura calculosorum" and "Aqua praeservans a calculo" von Johann Venod de Veteri Castro in Prag. Arch Gesch Med 7:396–402.
45. Forbes RJ (1970) A Short History of the Art of Distillation. E. J. Brill, Leiden, pp 108–112.
46. Inde AJ (1964) The Development of Modern Chemistry. Harper & Row, New York, pp 13–18.
47. Forbes RJ (1970) A Short History of the Art of Distillation. E. J. Brill, Leiden, p 217.
48. Anderson RGW (2000) The Archaeology of Chemistry. In: in Holmes FL, Levere TH (eds.), Instruments and Experimentation in the History of Chemistry. MIT Press, Cambridge, MA.
49. Thompson DJS (2002) Alchemy and Alchemists. Dover Publications, Inc., New York, pp 134-135.
50. al-Hassan AY (2009) Alcohol and the distillation of wine in Arabic sources from the 8th century. In Studies in Al-Kimiya': Critical Issues in Latin and Arabic Alchemy and Chemistry. Georg Olms Verlag, Hildesheim, Chapter 9.
51. Partington, JR (1998) A History of Chemistry, Martino Publishing, Mansfield Centre, CT, Vol. 2, pp 80–84.
52. Tubbs RS, Bosmia AN, Mortazavi MM, Loukas M, Shoja M, Gadol AAC (2012) Hieronymus Brunschwig (c. 1450–1513): his life and contributions to surgery. Childs Nerv Syst 28:629–632.
53. Forbes RJ (1970) A Short History of the Art of Distillation. E. J. Brill, Leiden, pp 120–124, 175–177.
54. Forbes RJ (1970) A Short History of the Art of Distillation. E. J. Brill, Leiden, p 32.
55. Forbes RJ (1970) A Short History of the Art of Distillation. E. J. Brill, Leiden, p 48.
56. Taylor FS (1992) The Alchemists. Barnes & Noble, New York, pp 94–100.
57. Crosland MP (1978) Historical Studies in the Language of Chemistry. Dover Publications, Inc., New York, pp 285–286.
58. Read J (1995) From Alchemy to Chemistry. Dover, New York, p 21.
59. Fleming A (1975) Alcohol, the Delightful Poison. Delacorte Press, New York, p 12.
60. Thonpson DJS (2002) Alchemy and Alchemists. Dover Publications, Inc., New York, pp 79–83.
61. Sarton G (1931) Introduction to the History of Science. The William & Wilkins Co., Baltimore, MD, Vol. II, part II, pp 893–900.
62. Benton JF (1982) The Birthplace of Arnau de Vilanova: A case for Villanueva de Jilóca near Daroca. Viator 13:245–257.
63. Hart E (1897) Archieologica Medica. XXIX. - Arnald of Villanova, The Mystic Physician. Brit Med J 1:1001.
64. Newman WR (2004) Promethean Ambitions. Alchemy and the Quest to Perfect Nature. University of Chicago Press, Chicago, pp 89–90.

Chapter 7
Early Chemical and Medical Applications of Alcohol

Alcohol had a profound influence on the art of the apothecary or pharmaceutical chemist [1], as well as changing the way that many artisans and alchemists carried out some types of reactions in the process of chemical pursuits. In this final chapter, the use of alcohol in early chemical and medical applications will be discussed, first as a simple component of fermented beverages and later in its purified form. In the process, the lasting effects of the availability of alcohol on the practice of both medicine and the chemical arts will be addressed.

7.1 Chemical and Medical Uses of Fermented Beverages

Early applications of fermented beverages as either medicines or chemical reagents most frequently specify wine (presumably grape wine), although the use of beer, date wine, palm wine, and drinks of fermented honey have all been reported as well. In terms of medicines, Babylonian and Assyrian authors mention the use of both beer and wine for such applications [2]. In the simplest sense, such drinks were considered to be nutritionally beneficial [3]. For this reason, physicians such as Hippocrates ($\sim$460–377 BCE)[1] and his followers commonly prescribed wine as a curative for some ailments by allowing the wine to strengthen the body [3–5]. In addition, Hippocrates also applied wine to quell fevers, as well as using it as a purgative and a diuretic [3]. The first century BCE Roman physician Asclepiades

---

[1] The Greek philosopher Hippocrates of Cos was reputed to be the greatest doctor of his time [10]. Although he is mentioned in the writings of Plato and Aristotle, very little is actually known about the man himself [11]. About 60 books have been ascribed to him, although the extent of the Hippocratic Corpus believed to be the work of Hippocrates himself remains uncertain [10, 11]. Hippocrates is often seen as the founder of scientific medicine [9] and some believe his writings constitute the first written expression of scientific thought [12]. The writings ascribed to him reveal a strong belief in the healing power of nature and suggest the primary role of the physician was to relieve pain and strengthen the patient's body and spirit, thus allowing nature to assert itself and function without hindrance to reestablish the body's equilibrium.

was noted for his liberal use of wine to treat patients, although his opponents viewed this simply as an attempt to indulge his wealthy patients [6].

Wine was also applied externally and was used by surgeons in washing wounds from antiquity until the 19th century [3, 7, 8]. The earliest known mention of washing wounds with fermented beverages, however, is found in a Sumerian clay tablet dated to 2100 BCE which describes the washing of wounds with beer and hot water [8]. Hippocrates prescribed wine for the dressing of wounds [3], as illustrated by the following instructions [9]:

> …the part is to be covered with unscoured wool, which is to be sprinkled from above with tepid wine and oil, but on no account is either bandage or compress to be applied; for this should be known most especially, that whatever compresses, or is heavy, does mischief in such cases. And certain of the dressings used to recent wounds are suitable in such cases; and wool may be laid upon the sore, and sprinkled with wine, and allowed to remain for a considerable time.

The treatment of wounds with wine is also mentioned in *The Bible* during the parable of the Good Samaritan [13]:

> He went to him and bandaged his wounds, pouring on oil and wine. Then he put the man on his own donkey, took him to an inn and took care of him.

In addition, it has been reported that for the common practice of bleeding a patient, swabbing the area to be cut with wine was particularly recommended by ancient physicians [14]. Although it is now known that wine is more bacteriocidal than 10 % alcohol [8], claims by some [3] that wine's antiseptic properties were understood in antiquity seem doubtful.

Drinks of fermented honey (hydromel, mulsum and oxymel) were also recommended in the 4th century BCE for treating many disorders [15]. Such treatments were popular with the Ionian medical school at Cnidus (or Knidos, west of Marmaris in present-day Turkey) [14] and are later described by Pliny the Elder in his *Naturalis Historia* [16]:

> Hydromel is recommended, too, as very good for a cough: taken warm, it promotes vomiting. With the addition of oil it counteracts the poison of white lead; of henbane, also, and of the halicacabum, as already stated, if taken in milk, asses' milk in particular. It is used as an injection for diseases of the ears, and in cases of fistula of the generative organs. With crumb of bread it is applied as a poultice to the uterus, as also to tumours suddenly formed, sprains, and all affections which require soothing applications. The more recent writers have condemned the use of fermented hydromel, as being not so harmless as water, and less strengthening than wine. After it has been kept a considerable time, it becomes transformed into a wine, which, it is universally agreed, is extremely prejudicial to the stomach, and injurious to the nerves.

Apart from direct consumption, fermented drinks such as beer and wine were also widely used as an important medium for dissolving and dispensing medicinal compounds [3, 17, 18]. Biomolecular archaeological evidence for plant additives in fermented beverages dates from the early Neolithic period [18]. Later Babylonian and Assyrian authors mention the practice of mixing plants in wine for the production of medicines [2] and the preparation of medicines by dissolving powdered components in beer are given in the Ebers Papyrus[2] [19]. The preparation of medicines made with honey and beer, sweet beer, and date wine are also described in the Berlin Papyrus[3] [20]. The later writings of Hippocrates include a number of medicines prepared by mixing various components in wine, as illustrated by the following preparation for a liniment used to treat ulcers [21]:

> The dried gall of an ox, the finest honey, white wine, in which the shavings of the lotus have been boiled, frankincense, of myrrh an equal part, of saffron an equal part, the flowers of copper, in like manner of liquids, the greatest proportion of wine, next of honey, and least of the gall.

In such applications, the alcohol content in the various fermented drinks would better solubilize the various organic ingredients in comparison to simple water. Wine appeared to be the most commonly applied for such medications, which could be a result of the higher alcohol content typical of wine in comparison to beer, thus making it a better solvent for dissolving or extracting the active species from various plant species. As a result, wine and various medicinal wines resulting from such mixtures with other components became regular stock in apothecaries of the Middle Ages [22].

Fermented beverages were additionally applied as solvents for the preparation of a number of other chemical mixtures as well. In both Egypt and Persia, wine and palm wine were used as a component of perfumes, quite likely as a solvent for the extraction of essential oils from plants [23, 24]. In the same way, wine has been used as a solvent for the production of writing ink [25]. Lastly, wine was known to be used by the Egyptians as a component of the incense Kyphi, which in turn was sometimes used to spice wines [7].

A final frequently cited application of palm wine was in embalming during mummification [6, 26, 27]. The use of palm wine by the Egyptians in embalming has been discussed by the Greek historian Herodotus and in greater detail by later historian Diodorus Siculus [26]. In addition, its use has been reported to be confirmed by the detection of alcohol traces in the tissues of mummies [6, 27]. It is known that both palm wine and Phoenician wine were used in the washing of the body during embalming, but the suggestion that palm wine was used as an alcoholic solvent for resins used in the embalming process is wrong, as the commonly applied cedar and terebinth (*Pistacio terebinthus*) resins do not readily dissolve in strong wine [7].

---

[2] The Ebers Papyrus, or Papyrus Ebers, is an 108-page Egyptian medical papyrus written around 1550–1700 BCE [8, 18]. However, it is believed that most of the information was copied from earlier texts, perhaps dating as far back as 2640 BCE [8].

[3] The Berlin Papyrus is an ancient Egyptian papyrus from the Middle Kingdom (2050–1650 BCE).

7.2 Early Chemical Applications of Alcohol

After the isolation of alcohol from fermented beverages in the 12th and 13th centuries, such alcohol distillates became common reagents of the laboratory where they were used as a new type of powerful solvent [28]. Not only could alcohol solubilize most salts and other water-soluble substances, but it also dissolved many organic materials not soluble in water, such as fats, resins, and essential oils. In that respect, it was the first broadly applied solvent that could be used to solubilize less polar species. For example, it was the first liquid that could be used to extract the volatile aromatic substances from plants [29]. The Italian physician Taddeo Alderotti noted the capacity of alcohol to absorb the flavors of fruits, herbs, and spices steeped in it, and published several recipes for cordials [30]. The glassmaker Antonio Neri later reported the use of alcohol for the extraction of organic dyes from herbs and flowers as illustrated by the following [31]:

> Take of whatoever Herb, or Flower, of whatsoever colour you will, which being bruised green upon a leaf of white Paper, tinges it with it's colour, these are good, but the Herbs and Flowers which do not so, are not good, then put into a glass body ordinary Aqua vitae, the head must be as large as possible, and in the top thereof put the leaves of whatsoever Flower or Herbs, from which you would draw a tincture, the lute the joynts of the head, and thereto fit a receiver, then give a temperature heat, that the thinner parts of the Aqua vitae ascending to the head, and falling upon the leaves and Flowers, may suck out the tincture, and distill thence into the receiver coloured Red, and full of the tincture of the Flowers, making all the subtle part of the Aqua vitae to ascend so long as it comes coloured, and then distill this Aqua vitae coloured in a glass vessel, which will come over white, and may serve at other times, and the tincture will remain at the bottom, which must not be dried too much but moderately, and thus you shall have the tincture or Lake from all Flowers, and Herb, singular for Painters.

Regardless of the new availability of alcohol, however, wines and vinegars were still preferred for the extraction of herbs in the preparation of medicines.

The applications of alcohol as a solvent also had profound impact on the pursuit of alchemy. In particular, some viewed *aqua ardens* and *aqua vitae* as the end to the search for a solvent for the Philosopher's stone, in the preparation of the Elixir of Life [32]. Arnald of Villanova in particular is thought to be perhaps the first to see in *aqua ardens* a key to the preparation of the Philosopher's stone [28]. John of Rupescissa went on to extinguish heated gold leaf in alcohol [33, 34], which he called *"fixing the sun in our sky"*. This subsequently became a leading procedure for the preparation of potable gold, although by no means the only one [33]. Even in terms of more mundane functions, the ability to dissolve a wider array of chemical species via alcoholic distillates greatly expanded the number of possible useful solutions available to the practicing alchemist and in many ways changed the focus of chemical investigations.

Fig. 7.1 Engraving of Taddeo Alderotti of Florence (ca. 1210–1295) (The National Library of Medicine)

7.3 Early Medical Applications of Alcohol

By the mid-13th century, alcoholic distillates began to be used as a medicine [28, 35] and by the 14th century, both *aqua ardens* and *aqua vitae* had become not only important in medicine, but had begun to be applied to techniques for the general preservation of organic substances [34]. The Italian physician Taddeo Alderotti (Fig. 7.1) and the Franciscan Vitalis de Furno (ca. 1260–1327)[4] have been credited with the earliest application of medicinal alcohol [34–36]. The English physician Gilbertus Anglicus has also been reported to have recommended *aqua vitae* to strengthen travelers around 1250 CE [34].

---

[4] Known also as Vitalis du Four, Vital du Four, Vital du Fourca, and Joannes Vitalis. He was a Franciscan theologian and scholastic philosopher [28] who played a prominent role in the controversy over the Franciscan conception of *usus pauper* (i.e. the restricted use of goods) [37]. He was born at Bazas in Aquitaine around 1260, about 60 km southeast of Bordeaux [37]. He entered the Franciscan order at an early age and went to study theology at Paris from 1285 to 1291. He taught at Montpellier from 1292 to 1296, after which he was transferred to the University of Toulouse [37]. He was made cardinal-priest by Pope Clement V in 1312 [28, 37] and became bishop of Albano in 1321. He died at Avignon on August 16, 1327 [28].

As both a physician and the inventor of the first method for the efficient production of alcoholic distillates, it is not surprising that Alderotti was one of the pioneers of the application of these distillates to medicine. He discusses the distilling of alcohol and wrote on its medicinal value in his *De virtutibus aquae vitae*, the final section of his *Consilia Medicinalia* written about 1280 CE [28, 30, 34, 38]. In these writings, Alderotti described how he used the four times redistilled wine for medical purposes. The ten-times distilled, obtained only in small quantities, he called *perfectissima*. It would burn completely away when ignited, and moreover any cloth soaked in it would burn also. Von Lippman quotes Alderotti's detailed description of his procedure and calculated that at least 90 % alcohol could be obtained by such fractional distillation [28, 34].

On the subject of alcohol, Alderotti abandoned his usual reserve in order to endorse *aqua vitae* as a medicinal substance in the warmest possible terms [30]. According to Alderotti, *aqua vitae* as a medicine was "*of inestimable glory, the mother and mistress of all medicine*". He went on to remark on its ability to fight melancholy, reporting that a little every morning "*makes one happy, jocund,*[5] *and glad*", and discussed its usefulness as a toothache liniment and for cleaning wounds [30]. Alderotti attributed a number of additional virtues to alcohol as a medicine, including the power to restore lost memory, strengthen weak sight, and relieve paralysis of the limbs, as well as being effective in the treatment of epilepsy and deafness [30, 38].

While less influential, Vitalis de Furno's *Pro conservando sanitate* described the preparation of *aqua ardens* from mixtures of good red wine and powdered sulfur. He then continues to discuss *aqua ardens* as not only a good medicine, but also a most useful solvent [28]. The application of alcohol as a medicine generally stemmed from the reasoning that purified alcohol would in turn purify the patient from illness and by 1,288 medicinal alcohol was in general use [29]. Its effect on patients that consumed it could be clearly seen. Administering medicinal alcohol resulted in relieved pain, improved mood, and made patients relaxed, thus allowing the body a chance to heal itself [39]. In addition, its effect on the failing powers of the elderly led to the use of alcohol distillates as a medicine against the ills of old age [29].

Aqua ardens and *aqua vitae* were thought to be able to preserve the human body, a belief that was supported by the fact that such alcohol distillates exhibited the property of preserving organic matter from putrefaction [29]. In a more practical sense, however, it is now understood that washing wounds with alcohol cleansed them and killed some microorganisms that could lead to infection. Alcohols exhibit broad-spectrum antimicrobial activity against most bacteria, fungi, and many viruses, but are ineffective against bacterial spores. Generally, the antimicrobial activity of alcohols is significantly lower at concentrations below 50 % and is optimal in the 60–90 % range [40]. As such, *aqua ardens* would have been more effective than the more concentrated *aqua vitae*. Little is known about

---

[5] Cheerful or merry.

7.3 Early Medical Applications of Alcohol

the specific mode of antimicrobial action of alcohols, but it is generally believed that they cause membrane damage and rapid denaturation of proteins [40].

In 1347–1351, Europe was in crisis as it dealt with an epidemic of plague, the well-known Black Death, and alcohol was the primary medicine that could give relief [35, 39, 41, 42]. As a cure for the plague, alcohol was ineffective, but at least it made the patient who drank it feel more robust. No other known treatment could even do that much [39]. Alcohol was also prescribed for cases of typhoid fever, diarrhea, and other similar diseases [43].

By the mid-14th century, the medicinal and preservative properties of 'pure' alcohol became the backbone of the writings of such authors as Arnald of Villanova (Fig. 7.1) and John of Rupescissa,[6] and it was soon widely recommended as a universal remedy [27, 34]. Arnald of Villanova describes alcoholic distillates as early as 1309–1312 and extols the virtues of *aqua ardens* in his *Liber de Vinis*, being one of the early authors to insist upon its curative virtues as a medicine [28, 29, 35]. Both in *Liber de Vinis* and its sequel *Tractatus de aquis medicinalibus*, he describes pharmaceutical wines and the distillation of wine with spices and sugar. For example, he describes a distillate of wine and rosemary that later became popular and was known as *aqua Hungarica* [28]. As such, he has been credited with the introduction of alcohol, medicinal wines, and related tinctures into the *Pharmacopecia* [44].

The principle work of John of Rupescissa was his *De consideratione quintae essentiae* (On the consideration of the fifth essence) (Fig. 7.1), which was written in the first half of the 14th century CE [28, 34]. This work consisted of two parts – canons and remedies, the backbone of which seems to be a presentation of the medicinal and preservative properties of alcohol [28] and a praise of its medical efficacy [33]. In this work he states that since a body cannot be preserved from corruption by things which are themselves corruptible, a remedy must be sought which is related to the four qualities as heaven is related to the four elements [33]. He conceives that 'pure' alcohol is the desired supreme remedy against corruption and thus gives it the fitting name of quintessence [28, 33]. He stated that the desired quintessence could be identified by its marvelous odor, quite different from that of simple *aqua ardens* [28]. He viewed this quintessence as a stabilizer which would defend the body from corruption indefinitely. This idea had some legitimacy and may well have derived from actual observations of the preservation of perishable plant and animal matter in alcohol [34].

---

[6] John de Rupescissa or John of Roquetallaide (d. 1362). An often cited author about which little is really known, he lived in the middle of the 14th century and was a tertiary member of the Franciscan order [28, 33]. He was known to his contemporaries for his apocalyptic preaching, for which he was often imprisoned [28, 33]. He was a Catalan, but a significant number of his books were written in Latin [28]. He studied in Toulouse for 5 years [28] before entering the Franciscan monastery at Orléans, where he continued his studies for 5 more years. He was imprisoned for the first time in 1345 [33] or 1346 [28], and again in 1349 and 1356 [28, 33]. His principle work was his *De consideratione quintae essentiae*, the backbone of which seems to be the medicinal and preservative properties of 'pure' alcohol [28].

Fig. 7.2 Engraving of Arnald of Villanova (ca. 1240–1311) from a 1682 manuscript

ARNALDVS — VILLANOVANVS —

The idea of associating alcohol with the Aristotelian fifth element is frequently credited to Rupescissa, for such earlier writers as Alderotti, Arnald of Villanova, and de Furno seem to have referred to alcohol only as either *aqua vitae* or *aqua ardens* [33]. Others, however, have given this credit to Arnald of Villanova [44] and works attributed to Ramon Lull have also been pointed to as the first to treat *aqua vitae* as an impure form of the quintessence [29]. Nevertheless, the belief that alcohol was the quintessence gave reason for the presumption that it would prove to be the most perfect of medicines. In addition, alcohol as the quintessence would then became a link between mankind's earthly bodies and the heavens, thus leading to the belief that it could transmit to mankind the beneficent influence of the heavens [29].

If the various claims for the power of alcohol above were not enough, the 15th century German physician Hieronymus Brunschwig later gave the following virtues in his *Liber deArte Distillandi: De Simplicibus* [45]:

> Aqua vitae is commonly called the mistress of all medicines. It eases the diseases coming of cold. It comforts the heart. It heals all old and new sores on the head. It causes a good color in a person. It heals baldness and causes the hair well to grow, and kills lice and fleas. It cures lethargy. Cotton wet in the same time and a little wrung out again and so put in the ears at night going to bed, and a little drunk thereof, is of good against all deafness. It eases the pain in the teeth, and causes sweet breath. It heals the canker in the mouth, in the teeth, in the lips, and in the tongue. It causes the heavy tongue to become light and well-speaking. It heals the short breath. It causes good digestion and appetite for to eat, and takes away all belching. It draws the wind out of the body. It eases the yellow jaundice, the dropsy, the gout, the pain in the breasts when they be swollen, and heals all

Fig. 7.3 Title page of a 1597 Italian copy of John of Rupescissa's *De consideratione quintae essentiae*

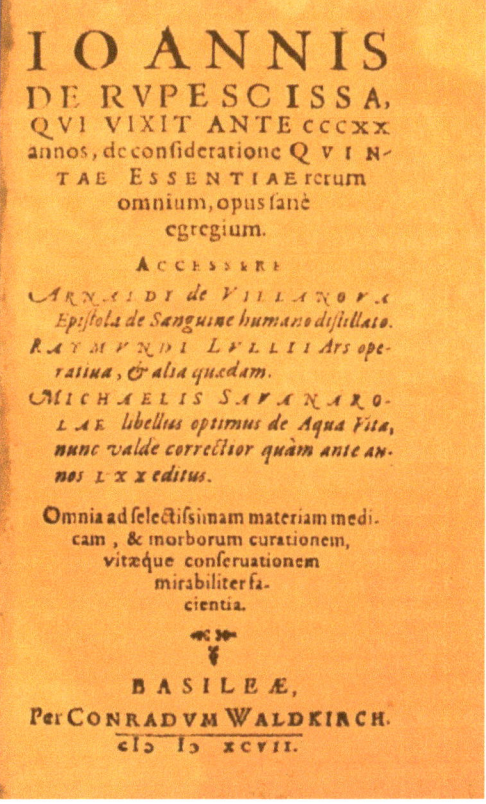

diseases in the bladder, and breaks the stone. It withdraws venom that has been taken in meat or in drink, when a little treacle is put thereto. It heals all shrunken sinews, and causes them to become soft and right. It heals the fevers tertian and quartan. It heals the bites of a mad dog, and all stinking wounds, when they be washed therewith. It gives also young courage in a person, and causes him to have a good memory. It purifies the five wits of melancholy and of all uncleanness.

Needless to say, alcohol was now an important aspect of medicine and well-equipped apothecaries of the Middle Ages had distillation equipment onsite for the production of *aqua ardens* and *aqua vitae*. Some inventories of this time period list quantities of these alcohol distillates in the tens of liters, although most such documents give smaller amounts [22].

References

1. Forbes RJ (1970) A Short History of the Art of Distillation. E. J. Brill, Leiden, p 57.
2. Partington, JR (1935) Origins and Development of Applied Chemistry. Longmans, Green and Co., London, pp 314–315.

3. Murray MA, Boulton N, Heron C (2000) Viticulture and wine production. In Nicholson PT, Shaw I (eds) Cambridge University Press, Cambridge, pp 577–608.
4. Sarton, G (1952) A History of Science. Ancient Science through the Golden Age of Greece. John Wiley & Sons, Inc., New York, p 384.
5. Sarton, G (1970) A History of Science, Volume 1, Ancient Science through the Golden Age of Greece. W. W. Norton & Co., New York, p 343.
6. Nutton V (2013) Ancient Medicine, 2nd ed. Routledge, New York, pp 170–173.
7. Partington, JR (1935) Origins and Development of Applied Chemistry. Longmans, Green and Co., London, pp 169–174.
8. Broughton G II, Janis JE, Attinger CE (2006) A Brief History of Wound Care. Plast Reconstr Surg 117 (Suppl.):6S–11S.
9. Hippocrates (1849) On the Articulations. Adams F (trans) The Internet Classics Archive: Daniel C. Stevenson, Web Atomics, Part 63.
10. Meadows J (1992) The Great Scientists, Oxford University Press, Oxford, pp 12–13.
11. Nutton V (2013) Ancient Medicine, 2nd ed. Routledge, New York, pp 53–70.
12. Singer C (1997) A Short History of Science, to the Nineteenth Century. Dover Publications: New York, p 39.
13. Barker K (ed) (1995) The NIV Study Bible, 10th ann ed, Luke, 10:34.
14. Nutton V (2013) Ancient Medicine, 2nd ed. Routledge, New York, p 93.
15. Crane E (1999) The World History of Beekeeping and Honey Hunting. Routledge, New York, pp 514.
16. Pliny the Elder (1855) The Natural History. Bostock J, Riley HT (trans) Taylor and Francis, London, Book XXII, Chapter 52.
17. Barnard H, Dooley AN, Areshian G, Gasparyan B, Faull KF (2011) Chemical evidence for wine production around 4000 BCE in the Late Chalcolithic Near Eastern highlands. J Archaeological Sci 38:977–984.
18. McGovern PE, Mirzoian A, Hall GR (2009) Proc Natl Acad Sci USA 106:7361–7366.
19. Partington, JR (1935) Origins and Development of Applied Chemistry. Longmans, Green and Co., London, pp 184–185.
20. Partington, JR (1935) Origins and Development of Applied Chemistry. Longmans, Green and Co., London, p 193.
21. Hippocrates (1849) On Ulcers. Adams F (trans) The Internet Classics Archive: Daniel C. Stevenson, Web Atomics, Part 5.
22. Bénézet JP (2001) Vin et alcool dans les apothicaireries médiévales des pays du Sud. Rev Hist Pharm, 89:477–488.
23. Partington, JR (1935) Origins and Development of Applied Chemistry. Longmans, Green and Co., London, p 158.
24. Partington, JR (1935) Origins and Development of Applied Chemistry. Longmans, Green and Co., London, p 424.
25. Partington, JR (1935) Origins and Development of Applied Chemistry. Longmans, Green and Co., London, p 207.
26. Broshi M (2007) Date Beer and Date Wine in Antiquity. Palest Explor Q 139:55–59.
27. Partington, JR (1935) Origins and Development of Applied Chemistry. Longmans, Green and Co., London, pp 197–198.
28. Forbes RJ (1970) A Short History of the Art of Distillation. E. J. Brill, Leiden, pp 60–65.
29. Taylor FS (1992) The Alchemists. Barnes & Noble, New York, pp 98–100.
30. Siraisi NG (1981) Taddeo Alderotti and His Pupils. Two Generations of Italian Medical Learning. Princeton University Press, Princeton, pp 300–301.
31. Neri A, Merrett C (2003) The World's Most Famous Book on Glassmaking, The Art of Glass. The Society of Glass Technology, Sheffield, pp 221–222.
32. Read J (1995) From Alchemy to Chemistry. Dover, New York, p 21.
33. Multhauf JP (1954) John of Rupescissa and the Origin of Medical Chemistry. Isis 45:359–367.

References

34. Gwei-Djen L, Needham J, Needham D (1972) The Coming of Ardent Water. Ambix 19:69–112.
35. von Lippmann EO (1912) Zur Geschichte des Alkohols und seines Namens. Angew Chem 40:2061–2065.
36. Stillman JM (1924) The Story of Early Chemistry. D. Appleton and Co., New York, pp 187–192.
37. Traver AG (2003) Vital du Four. In: Gracia JJE, Noone TB (eds.) A Companion to Philosophy in the Middle Ages. Blackwell Publishing Ltd, Malden, pp 670–671.
38. von Lippmann EO (1914) Thaddäus Florentinus (Taddeo Alderotti) über den Weingeist. Arch Gesch Med 7:379–389.
39. Vallee BL (1998) Alcohol in the Western World. Sci Am 279(June):80–85.
40. McDonnell G, Russell AD (1999) Antiseptics and Disinfectants: Activity, Action, and Resistance. Clin Microbiol Rev 12:147–179.
41. Forbes RJ (1970) A Short History of the Art of Distillation. E. J. Brill, Leiden, p 91.
42. Liebmann, AJ (1956) History of Distillation. J Chem Educ 33:166–173.
43. Forbes RJ (1970) A Short History of the Art of Distillation. E. J. Brill, Leiden, p 95.
44. Hart E (1897) Archieologica Medica. XXIX. - Arnald of Villanova, The Mystic Physician. Brit Med J 1:1001.
45. Roueché B (1963) Alcohol in Human Culture. In: Lucia SP (ed.) Alcohol and Civilization. McGraw-Hill, New York, pp 167–182.

About the Author

Seth C. Rasmussen is a Professor of Chemistry at North Dakota State University (NDSU) in Fargo (seth.rasmussen@ndsu.edu). He received his B.S in Chemistry from Washington State University in 1990 and his Ph.D. in Inorganic Chemistry from Clemson University in 1994, under the guidance of Prof. John D. Peterson. As a postdoctoral associate at the University of Oregon, he then studied conjugated organic polymers under Prof. James E. Hutchison. In 1997, he accepted a teaching position at the University of Oregon, before moving to join the faculty at NDSU in 1999.

Active in the fields of materials chemistry and the history of chemistry, his research interests include the design and synthesis of conjugated materials, photovoltaics (solar cells), organic light emitting diodes, the application of history to chemical education, the history of materials, and chemical technology in antiquity. As both author and editor, Prof. Rasmussen has contributed to books in both materials and history and has published more than 65 research papers and book chapters. He is a member of various international professional societies including the American Chemical Society, Materials Research Society, Alpha Chi Sigma, American Nano Society, Society for the History of Alchemy and Chemistry, History of Science Society, and the International History, Philosophy and Science Teaching Group.

Prof. Rasmussen currently serves as the Program Chair for the History of Chemistry division of the American Chemical Society, as a member of the Editorial Board for the journal *Topological and Supramolecular Polymer Science*, and as Series Editor for *Springer Briefs in Molecular Science: History of Chemistry*.

Index

A
Abydos, 42, 54
Acetic acid, 14, 66, 72
Alcohol, 2, 3, 6, 7, 9, 14, 21, 24, 25, 29, 30, 33, 40, 45, 66, 72, 76, 79, 82, 83, 86, 88–92, 95–103
Alderotti, Taddeo, 84
Alembic, 83, 84, 86, 89, 90
Amarna, 45, 54
Ambix, 80, 83
Amphora, 62
Anatolia, 9, 34, 41, 55, 59
Antimony trisulphide, 4
Aqua ardens, 89, 91, 92, 98–103
Aqua flamens, 91, 92
Aqua vitae, 91, 92, 98–103
Armenia, 56
Arnald of Villanova, 92, 98, 101, 102

B
β-galactosidase, 73
Babylonia, 22
Bacteria, 8, 66, 71, 72, 100
Bag press, 64
Bappir, 39, 41
Barley, 8, 23, 24, 32–36, 40–45, 56
Beer, 7–9, 13, 17, 18, 20–23, 30–46, 62, 71, 95–97
Beerstone, 20, 37, 38
Beijerinck, Martinus Willem, 73
Berzelius, Jöns Jacob, 3, 5
Bikos, 80
Biomarker, 19, 20, 51, 54–57
Black Death, 101
Bread, 8, 18, 30–33, 35, 42, 44–46, 96
Brewing, 31, 32, 35, 39–42, 44–46
Bronze Age, 36, 51, 60, 67
Brunschwig, Hieronymus, 88, 102
Buchner, Eduard, 16

C
Calcium, 20, 55, 82, 84
Cetyl alcohol, 3
Chemical analysis, 18, 43, 51, 52, 57
China, 9, 18, 19
Clay, 66, 72, 82, 83, 90, 96
CO_2, 30, 72, 74, 76
Columella, 20, 21, 66
Cucurbit, 80, 82, 86, 90

D
Dalton, John, 2, 4
Date, 18, 22–25, 29, 32, 34, 41, 43, 46, 52, 71, 75, 79, 89, 95, 97
Date palm, 22–24, 32, 49
Date wine, 22–24, 34, 95
de Furno, Vitalis, 99, 100
Distillation, 1, 5, 9, 79, 80, 82–86, 88–90, 92, 100, 101, 103
Dumas, Jean Baptiste, 2, 4

E
Ebers Papyrus, 13, 97
Egypt, 4, 8, 22–25, 32, 34, 36, 37, 42, 43, 45, 60, 62, 67, 97
Embalming, 97
Ethanol, 1, 3, 7, 13, 17, 66, 72, 79, 89
Euphrates, 8, 36, 40

F
Feigl spot test, 19, 37, 38, 52, 53
Fermentation, 1, 7–9, 13, 14, 16–18, 20–22, 24, 25, 29, 30, 33, 35, 38, 41–43, 45, 46, 57, 64, 66, 71–75, 79
Fertile Crescent, 8, 9, 22, 29, 60
Fourier-transform infrared spectroscopy (FTIR), 18, 19

France, 2, 32, 35, 75
Fructose, 15, 66
Fruit, 7, 19, 22, 24, 43, 54, 55, 61, 63, 66

G
Galactose, 43, 72, 73, 75
Gas chromatography (GC), 43, 52
Gay-Lussac, Joseph, 14, 15
Gesner, Conrad, 88
Glass, 9, 80, 82–84, 86, 88, 90, 98
Glucose, 14, 15, 17, 29, 43, 66, 72–74
Goats, 8, 71
Godin Tepe, 37, 52
Grain, 7, 13, 14, 22, 29, 31, 32, 35–37, 39, 41, 45, 46
Grapes, 7, 13, 19, 35, 43, 49, 52, 54–59, 62, 63, 64, 67
Grape wine, 20, 34, 35, 49, 51, 62, 63, 95
Greece, 60, 62
Greek, 4, 35, 36, 42, 49, 60, 75, 97

H
Herodotus, 35, 75, 97
High-performance liquid chromatography (HPLC), 19, 52
Hippocrates, 95–97
Honey, 7, 13, 18–22, 39, 41, 45, 67, 95–97
Hops, 22, 33
Hydromel, 18, 20, 21, 96

I
Italy, 60, 89

J
John of Rupescissa, 98, 101
Jordan Valley, 60

K
Kefir, 71–73, 75, 76
Kefir grains, 72, 73, 75, 76
King Midas, 19
Kohl, 3–5
Kumis, 71, 75, 76

L
Lactic acid, 65, 71–74, 76
Lactose, 72, 73, 76
Lavoisier, Antoine, 14, 15

Levant, 37, 58, 60, 62, 83
Lull, Raymond, 91
Lute, 82, 83, 91, 98

M
Malt, 29, 33, 34, 40, 41, 45, 46
Maltose, 29, 32
Malvidin, 55–57
Maria the Jewess, 79, 80
Mass spectrometry (MS), 18
Mead, 13, 18–22, 35
Medicine, 95, 99–101, 103
Mesopotamia, 9, 22, 23, 32, 34, 36, 37, 41, 51, 54, 59, 60, 62, 67
Methanol, 3
Milk, 9, 13, 71, 72, 75, 76, 96
Mold, 57, 66, 72
Moor's head, 86, 88
Must, 44, 60, 64

N
Natron, 83
Near East, 29, 32, 34, 52, 55, 58, 63
Neolithic, 8, 18, 49, 52, 58, 96
Nile, 8, 24, 43, 45, 51, 60, 62
Ninkasi, 35, 39, 40
Noah Hypothesis, 58

O
Oxalate, 20, 37, 38

P
Paleolithic, 8, 18, 49
Palestine, 22, 25, 51, 62
Palm wine, 24
Paracelsus, 5, 6, 92
Peligot, Eugène, 2
Pliny the Elder, 20, 21, 23, 24, 66

Q
Quintessence, 91, 101, 102

R
Residues, 38, 43, 45, 46, 51, 52, 57
Resin, 1, 52, 54, 66, 67
Retort, 86, 87
Roman, 4, 20, 52, 62, 82, 83, 95
Rosenhut, 86, 88

S

Saccharomyces cerevisiae, 13, 21, 32, 35, 54, 65, 75
Salernus, 83, 89
Sheep, 8, 71, 72
Silica, 82
Soda, 82, 83
Solen, 80, 83, 84, 86, 90
Solvent, 1, 10, 56, 97, 98, 100
Starch, 29, 33, 46
Stibnite, 4
Sugar, 7, 8, 13, 14, 17, 18, 21, 22, 24, 29, 33, 43, 65, 72, 76, 101
Sumer, 37, 40
Syria, 9, 22, 56, 62
Syringic acid, 55, 56

T

Tartrate, 52, 54, 55, 90
Tartaric acid, 19, 52, 54, 55, 57
Terebinth, 52, 54, 66, 67, 97
Thenard, Louis, 2, 14
Tigris, 8, 36, 40
Transcaucasia, 50, 58, 59
Treading vat, 63
Turkey, 8, 19, 58, 96

U

Uruk, 37, 39, 52, 54

V

Viniculture, 49, 51, 57–60
Vitis vinifera sativa, 49
Vitis vinifera sylvestris, 49, 57, 58, 60
Vitis vinifera vinifera, 14, 16, 49
von Baeyer, Adolf, 14, 16
von Liebig, Justus, 30

W

Wedel, Georg, 6
Wenod, Johann, 85
Wheat, 8, 24, 32, 42, 43, 56
William of Rubruck, 75
Williamson, Alexander, 2, 3
Wine, 3, 6–8, 10, 13, 18–24, 32, 34–38, 40, 49, 51, 52, 54, 55, 57–60, 62, 64, 66, 67, 71, 76, 84, 88–91, 95–97, 100, 101
Wort, 33

X

Xenophon, 34, 35

Y

Yeast, 13, 16, 17, 24, 29, 32, 35, 45, 46, 57, 64, 66, 73

Z

Zagros, 37, 49, 52, 58, 60
Zosimus, 44
Zymase, 16, 17

MIX
Papier aus verantwortungsvollen Quellen
Paper from responsible sources
FSC® C105338

If you have any concerns about our products,
you can contact us on
ProductSafety@springernature.com

In case Publisher is established outside the EU,
the EU authorized representative is:
**Springer Nature Customer Service Center GmbH
Europaplatz 3, 69115 Heidelberg, Germany**

Printed by Libri Plureos GmbH
in Hamburg, Germany